Springer Theses

Recognizing Outstanding Ph.D. Research

Aims and Scope

The series "Springer Theses" brings together a selection of the very best Ph.D. theses from around the world and across the physical sciences. Nominated and endorsed by two recognized specialists, each published volume has been selected for its scientific excellence and the high impact of its contents for the pertinent field of research. For greater accessibility to non-specialists, the published versions include an extended introduction, as well as a foreword by the student's supervisor explaining the special relevance of the work for the field. As a whole, the series will provide a valuable resource both for newcomers to the research fields described, and for other scientists seeking detailed background information on special questions. Finally, it provides an accredited documentation of the valuable contributions made by today's younger generation of scientists.

Theses may be nominated for publication in this series by heads of department at internationally leading universities or institutes and should fulfill all of the following criteria

- They must be written in good English.
- The topic should fall within the confines of Chemistry, Physics, Earth Sciences, Engineering and related interdisciplinary fields such as Materials, Nanoscience, Chemical Engineering, Complex Systems and Biophysics.
- The work reported in the thesis must represent a significant scientific advance.
- If the thesis includes previously published material, permission to reproduce this must be gained from the respective copyright holder (a maximum 30% of the thesis should be a verbatim reproduction from the author's previous publications).
- They must have been examined and passed during the 12 months prior to nomination.
- Each thesis should include a foreword by the supervisor outlining the significance of its content.
- The theses should have a clearly defined structure including an introduction accessible to new PhD students and scientists not expert in the relevant field.

Indexed by zbMATH.

Patrick T. MacGregor

Single-Particle Structure of ^{29}Mg on the Approach to the $N = 20$ Island of Inversion

Doctoral Thesis accepted by
The University of Manchester, Manchester, UK

Springer

Author
Dr. Patrick T. MacGregor
The University of Manchester
Manchester, UK

Supervisor
Dr. David Sharp
The University of Manchester
Manchester, UK

ISSN 2190-5053 ISSN 2190-5061 (electronic)
Springer Theses
ISBN 978-3-031-19118-3 ISBN 978-3-031-19119-0 (eBook)
https://doi.org/10.1007/978-3-031-19119-0

This Springer imprint is published by the registered company Springer Nature Switzerland AG
The registered company address is: Gewerbestrasse 11, 6330 Cham, Switzerland

The Son is the image of the invisible God, the firstborn over all creation. For in him all things were created: things in heaven and on earth, visible and invisible, whether thrones or powers or rulers or authorities; all things have been created through him and for him. He is before all things, and in him all things hold together.

COLOSSIANS 1:15–17[1]

To my wife Ally, your unwavering support has helped me to produce this work and my love for you compels me to recommend other reading material, unless you've really run out of all other options.

Supervisor's Foreword

In recent years, it has been shown that the magic numbers in nuclei, corresponding to closed shells of neutrons or protons in the nuclear shell model, evolve with proton and neutron number. They are not as immutable as presented in undergraduate textbooks; some shell closures disappear, whilst others appear in nuclei away from the valley of β stability. To understand the mechanisms that drive this evolution of nuclear structure, physicists must study the structure of exotic nuclei, in particular, the characteristics of the individual nucleons within them, what is known as single-particle structure. Typically, this information is accessed by measuring direct nuclear reactions using accelerated beams of ions.

Direct reactions allow access to many nuclear properties. They are reactions that proceed via a single step, exciting just one degree of freedom in the residual nucleus. They can probe single-particle properties via the transfer of a single nucleon from a light ion to or from a nucleus of interest. Pairing properties can be investigated by transfer of a correlated pair and collective properties via inelastic scattering. To measure these reactions on short-lived nuclei, it is necessary to use "inverse kinematics" where the beam particle is the radioactive nucleus and the light ion is the target.

Radioactive ion beam facilities, such as ISOLDE at CERN, allow nuclear physicists access to exotic nuclei away from stability. Coupled to these facilities, the latest particle-detection techniques are required to study the structure of these nuclei via direct reactions in inverse kinematics. Solenoidal spectrometers are a novel device that offer the best charged-particle resolution compared to other techniques.

Patrick's thesis describes the first measurement made using a new solenoidal device, the ISOLDE Solenoidal Spectrometer at the ISOLDE facility. These measurements of the $d(^{28}\mathrm{Mg},p)^{29}\mathrm{Mg}$ reaction probe the evolution of nuclear structure along nuclei with $N = 17$ investigating the emergence of a new shell closure at $N = 16$, testing the modern effective shell model interactions, as well as illuminating the role that the finite geometry of the nuclear potential plays in observed systematics of single-particle energies. These data are important in constraining models in a rapidly evolving region of the nuclear chart. As well as the emergence of a new shell closure at $^{24}\mathrm{O}$ in lighter $N = 16$ nuclei, the heavier magnesium isotopes sit in a region of changing nuclear shape called the island of inversion, where the weakening of the $N = 20$ shell gap results in the onset of deformation in the nucleus. These phenomena are of topical interest in the field, and Patrick's findings will provide a key benchmark.

Patrick's thesis also provides a canonical reference for nuclear physicists on the use of solenoid spectrometers to study direct nuclear reactions in inverse kinematics and will provide a useful guide for years to come. He has covered in detail the kinematics of reactions in a solenoid field and the transformations required to extract the Q-value of the populated states in the reaction from the measured quantities.

May 2022

Dr. David Sharp
STFC Ernest Rutherford Fellow
The University of Manchester
Manchester, UK

The original version of this book inadvertently contained incorrect typesetting within Equations A.2, A.3, A.5, and A.9. This has now been rectified.

Abstract

The nuclear structure of ^{29}Mg was probed using the $d(^{28}\mathrm{Mg},p)^{29}\mathrm{Mg}$ transfer reaction to populate its single-particle states. The ISOLDE facility at CERN provided a 9.473 MeV/u beam of ^{28}Mg which was directed at a deuterated target within the ISOLDE solenoidal spectrometer. Exploiting the kinematic advantages of this technique allowed most states up to 5 MeV to be resolved and angular distributions of the reaction cross section to be obtained. The DWBA code DWUCK5 was used to obtain spectroscopic factors for these states. Additionally, some higher-lying excited states were identified, and their possible properties were proposed.

Theoretical calculations in this region broadly reproduced the observed behaviour in ^{29}Mg, as well as matching trends from other nuclides in the $N = 17$ isotones. These calculations indicate that the nucleon–nucleon interaction between protons and neutrons, with the tensor interaction as a key component, is driving the evolution of shell structure, with the filling of the $\pi d_{5/2}$ orbital of particular importance in this region. Finite geometry effects also play an important role in this evolution as the p orbitals approach the neutron-separation threshold and the energy spacing between them reduces.

The information gained from the nuclear structure will help to continue refining these interactions and provide a valuable benchmark for nuclear-structure studies around the border of the island of inversion. Similar experiments to study ^{30}Al and ^{31}Mg carried out recently will help to further the understanding of nuclear structure in this exotic region of the nuclear chart.

Acknowledgements

This work before you is the culmination of eight amazing years in Manchester, marking my progression from an inexperienced (and nerdy) teenager to a more competent (but still nerdy) adult. In that time, I have married, lost a lot of my hair and had a fantastic time learning about the wonderful world of physics. This journey has not been completed alone, and there are far too many people to thank.

My primary thanks must go to Prof. Sean Freeman and Dr. David Sharp, who have guided me through my Ph.D. with much patience and care. The depths of Sean's nuclear physics knowledge have been an invaluable resource for diagnosing my many mis-steps and helping me find further sources of information about a variety of topics. Despite his demanding duties as the head of department and now the head of ISOLDE, he has always found time to help me, for which I am immensely grateful. I hope my hyphenation and grammar have improved as a result of the many corrections you gave.

David has borne the brunt of my questions, poor writing and misunderstandings. His dedication to the field and to his students is commendable, and I have been the recipient of his wisdom and kindness. Despite juggling many responsibilities, he has always found time for me. We have enjoyed the slight delirium that accompanies night shifts on experiments more than once, and I hope we can continue to work together for a long time, though perhaps with better sleep in future.

Beth Cropper has been a constant source of bad jokes and another delightful person to share late-night shifts with on experiments. It has been a pleasure to get to know Katie Garrett, and it would have been lovely to work together in person more—I wish you both every success on the rest of your projects. I am also indebted to the wider Manchester nuclear physics group, past and present. Special thanks go to Stuart Szwec for helping me get started and to everyone who helped to turn our office into a mini-jungle and make it a fantastic place to work. In addition, my fellow student Sam Bennett and I have been together throughout our university experience, not only as housemates and laboratory partners, but also good friends. I am incredibly excited that you get to continue researching fission using ISS with Sean and David! (I should also confess that the peculiar behaviour of your disc drive was entirely down to me!)

Additional thanks go to the target group at Argonne National Laboratory and the team at ISOLDE, CERN who were able to deliver the beam and troubleshoot all of our issues whether convenient or not.

I could not have completed this project without the support of friends outside of nuclear physics. Special thanks to Amy, Ali, Lydia, Mike, Pawita and Parinya for your friendship (and amazing food!) throughout both good and bad times. Special thanks to Jack for fixing my laptop at pretty short notice, and to Carl for steadfastness, integrity and friendship—I am glad I can call you my friend.

My family cannot escape being mentioned in this discourse. I am indebted to my parents and grandparents, who have all sacrificed much ease and comfort to make me the man I am today. Stuart, my fellow physicist, twin and best friend has been an invaluable support to me for my whole life (literally). Matthew, Elizabeth, Angus and Cath have all supplied merciful distraction and relief often. 媽媽和爸爸: 多謝你們的愛和好意! 我好感激你們的支持. Abby, Ernest and Anna have welcomed me into their family with open arms and kindness, and I could not ask for better siblings-in-law.

However, my Ph.D. could not have been completed without my incredible wife, Ally. She has stood by me throughout this whole project, supporting me, challenging me, pushing me and being an all-round amazing person. I do not know how I could have finished this without her help through the numerous challenges that both Ph.D. and life have thrown at me.

Soli deo gloria.

The $d(^{28}\mathrm{Mg},p)$ measurements with ISS were supported by the UK Science and Technology Facilities Council (STFC), the U.S. Department of Energy, Office of Science, Office of Nuclear Physics, under Contract Number DE-AC02-06CH11357 (ANL), the European Union's Horizon 2020 Framework research and innovation programme under grant agreement no. 654002 (ENSAR2), the Marie Skłodowska-Curie grant agreement No. 665779, the Research Foundation Flanders (FWO, Belgium), and the European Research Council under the European Union's Seventh Framework Programme (FP7/2007-2013)/ERC grant agreement number 617156.

Abbreviations

CM	Centre of momentum
DWBA	Distorted-wave Born approximation
EBIS	Electron Beam Ion Source
ESPE	Effective single-particle energy
HELIOS	HELIcal Orbital Spectrometer
HIE-ISOLDE	High Intensity and Energy ISOLDE
HRS	High-resolution separator
IK	Inverse kinematics
IoI	Island of inversion
IPM	Independent particle model
ISOLDE	Isotope mass separator online (DE)
ISS	ISOLDE solenoidal spectrometer
KC	Kinematic compression
KS	Kinematic shift
LINAC	Linear accelerator
NK	Normal kinematics
PSB	Proton synchrotron booster
REX-ISOLDE	Radioactive beam EXperiment at ISOLDE
RF	Radio frequency

Chapter 1
Introduction

The understanding of matter has long been a pursuit of scientists throughout history. Each scientific discipline adds to the sum total of this understanding from different perspectives. The nuclear physicist adds to this picture by examining the nature of atomic nuclei. This characterisation has yielded technological development in areas such as nuclear power and medical treatments.

The discovery of the atomic nucleus occurred in the early 20th century when Rutherford, Geiger, and Marsden observed α-particles scattering at greater angles than expected [1, 2]. It was later realised that this phenomenon is explained by the atom having a small and dense core: the nucleus. This breakthrough then necessitated a revision of the understanding of atoms at the time.

Further investigation led to the discovery of protons and neutrons, the building blocks of nuclei. The number of these nucleons in atomic nuclei ranges from one to a few hundred. Therefore, there are too few nucleons to use statistical models to extract nuclear properties. Furthermore, calculating the properties of nuclei from first principles is difficult.

The nucleon-nucleon force is complicated, requiring much computational power to accurately model all of the interactions within. Nevertheless, in the last decade, there has been much progress in the calculations of nuclear properties using ab initio methods. For example, Stroberg et al. [3] were able to calculate separation energies for nuclei beyond the $Z = 20$ and $N = 40$ shell gaps.

While there is currently no standard model that describes all nuclei perfectly, one of the most useful theoretical frameworks is the shell model, where nuclei are represented as being filled with nucleons in discrete shells of differing total angular momentum, j; parity, π; and principal quantum number, n. These shells are then filled in accordance with the Pauli Exclusion Principle with differing values of the magnetic quantum number, m. The shell model is described in more detail in Chap. 2.

P. T. MacGregor, *Single-Particle Structure of* ^{29}*Mg on the Approach to the N=20 Island of Inversion*, Springer Theses, https://doi.org/10.1007/978-3-031-19119-0_1

This model was originally developed with insight gained from measurements of stable nuclei. A number of observables demonstrate the existence of these shells, such as the proton, neutron, two-proton, and two-neutron separation energies; the charge radii of nuclei; and so on.

This led to the establishment of the "magic numbers", which represent the shell closures in either the numbers of neutrons or protons, marked by their increased stability compared to their neighbours that are one nucleon or more away. However, this static picture of shell structure that has been established for stable nuclei has been shown to change away from stability, where advances in radioactive ion beam facilities are allowing access to more exotic systems.

1.1 Changing Magic Numbers

An example of a change in the shell structure established along the line of β-stability is the weakening of the $N = 20$ shell closure and the emergence of a new one at $N = 16$ in neutron-rich nuclei. This is evidenced by the measurement of the energy of the first excited 2^+ states in even-even isotopes in this region, with a particular focus on ^{24}O, which has been investigated in Refs. [4, 5], shown in Fig. 1.1.

A rise in the excitation energy of the lowest 2^+ state is indicative of a shell gap. Figure 1.1 shows this in carbon and oxygen isotopes at the familiar $N = 8$ shell gap, but also the appearance of a new $N = 16$ shell gap in neutron-rich oxygen, supporting the conclusion that ^{24}O is doubly magic. Additionally, other sub-shell closures are also possible: the calcium isotopes have suggested sub-shell closures at $N = 32$ and 34 (see, for example, Ref. [6]).

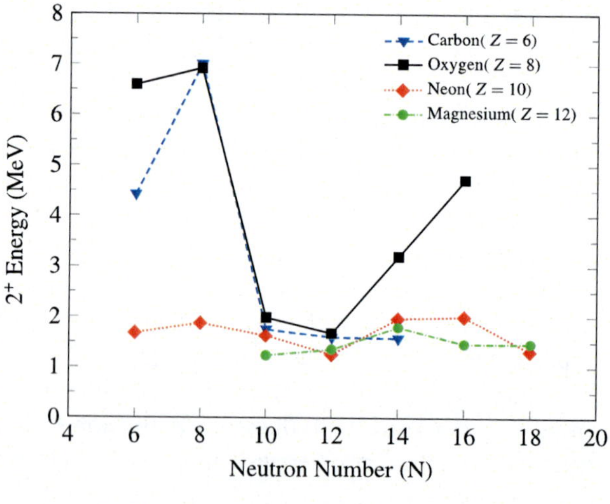

Fig. 1.1 The energy of the first excited 2^+ state in even-even isotopes of carbon, oxygen, neon, and magnesium. Adapted from Ref. [4]

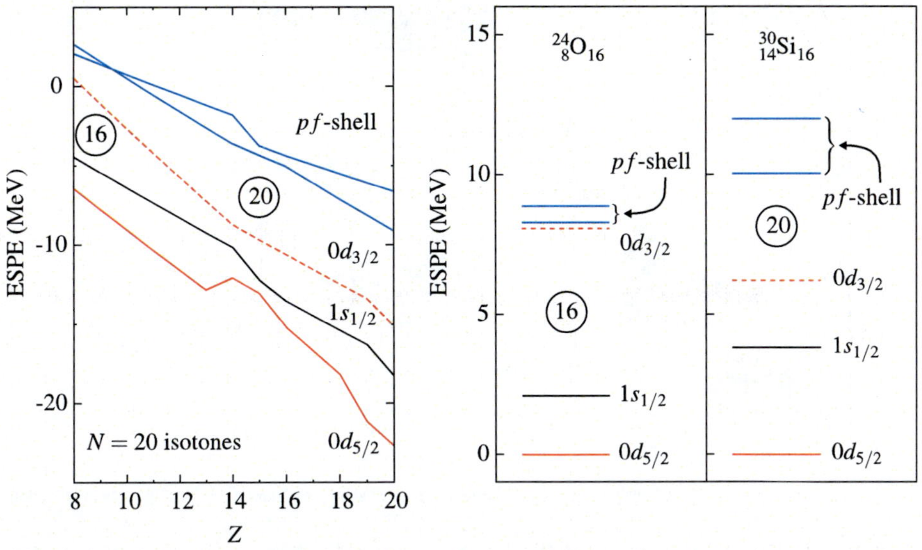

Fig. 1.2 Left: the change in ESPE with Z for the $N = 20$ isotones. Right: a comparison of the ESPE change between ^{24}O and ^{30}Si. Adapted from Ref. [7]

Otsuka et al. [7] attribute this evolution to changes in the interactions between valence protons and neutrons, as protons fill or are emptied from a particular orbital. Figure 1.2 shows the calculated effective single-particle energies (ESPEs) for the $N = 20$ isotones. Here, the calculations show that as Z decreases, the difference in ESPE between the $\nu d_{3/2}$ and the higher orbitals in the νpf-shell decreases as well.

This difference is attributed to the changing attraction between the $\nu d_{3/2}$ and $\pi d_{5/2}$ shells. As the $\pi d_{5/2}$ empties, the attraction between the two shells decreases, causing the $\nu d_{3/2}$ to rise in ESPE relative to the pf-orbitals. This effect is caused by the different monopole shift that each orbital experiences, with the greatest effect occurring between proton and neutron orbitals with large radial overlaps. These monopole shifts will be discussed further in Sect. 2.3.1.

As this shell gap weakens for neutron-rich nuclei, there is an increased probability of particle-hole excitations at lower excitation energies, changing the low-lying and ground-state configurations of nuclei, resulting in the $N = 20$ "island of inversion".

1.2 The Island of Inversion

In 1975, irregularities were discovered by Thibault et al. [8] in the binding energies of ^{31}Na and ^{32}Na. These nuclei were more bound than expected, and shell-model calculations in this region failed to reproduce this effect when constrained to the $1s0d$ shell. By allowing cross-shell interactions with the $0f1p$-shell, the trend in binding energies for the sodium isotopes was more successfully reproduced [9].

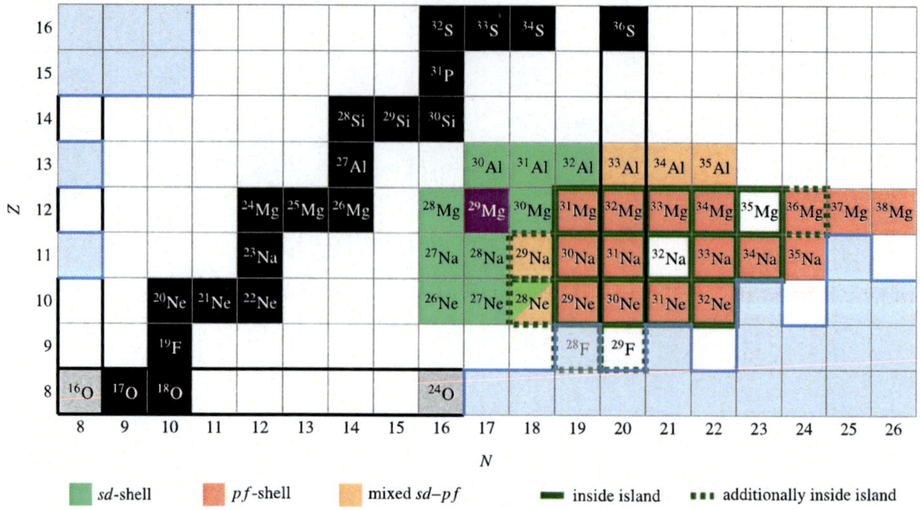

Fig. 1.3 A table of nuclides highlighting the $N = 20$ island of inversion. Colours listed represent ground-state configurations. Nuclei that decay by proton or neutron emission are highlighted in blue, doubly-magic nuclei are grey (including the previously mentioned ^{24}O), and ^{29}Mg has been highlighted in purple. Adapted primarily from Ref. [11], which gives nuclei within the island. Further information for this chart, including nuclei within the island that were not mentioned in Ref. [11], is from Refs. [9, 12–21]

This region of neutron-rich nuclei between fluorine and aluminium, where these effects are observed, is known as the $N = 20$ island of inversion (IoI).[1]

The current extent of the IoI is illustrated in the chart of nuclides in Fig. 1.3. As mentioned in the last section, the onset of this island occurs as negative-parity intruder states appear at low excitation, due to an increased probability of particle-hole excitations across the weakening $N = 20$ shell gap. These particle-hole excitations result in a prevalence of deformed configurations in the low-lying and ground states of nuclei in this region. Nuclei within the island have ground-state wave functions dominated by configurations consisting of fp-states, whilst those outside are dominated by sd wave functions.

The IoI was originally discovered, as mentioned, due to a discrepancy in binding energies of sodium isotopes. Hartree-Fock calculations, mass measurements, and further calculations helped to define the centre of the island at ^{32}Na [9]. Since then, the boundaries of the island have been established using different techniques. Kobayashi et al. [12] performed one-neutron removal from ^{29}Ne, indicating that it lies inside the island. Time-of-flight mass measurements by Gaudefroy et al. [13], and (p, pn) and $(p, 2p)$ reactions on ^{29}Ne and ^{29}F by Revel et al. [14] suggested that ^{28}F and ^{29}F lie inside the island, though they are unbound, and mark the southern border (low-Z) of the island.

---

[1] While a number of these islands have been discovered [10], the only relevant one for this work is at $N = 20$.

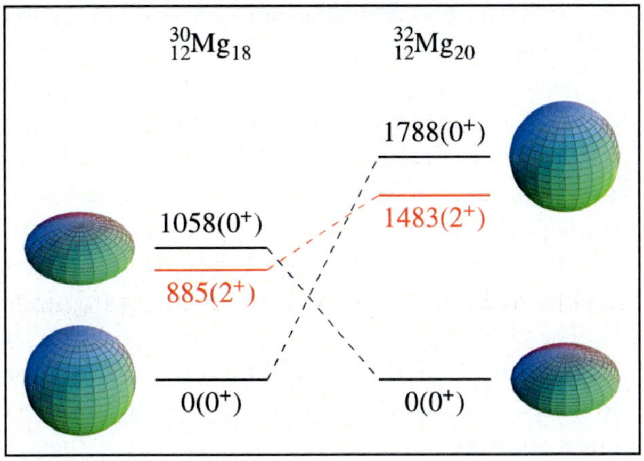

Fig. 1.4 The reordering of levels in isotopes of magnesium when transitioning into the IoI [22]

The west (low-N) part of the island was marked by Neyens et al. [15], who used collinear laser spectroscopy to probe the hyperfine structure of ^{31}Mg, and combined this with a β- detected nuclear magnetic resonance (β-NMR) measurement to extract the nuclear magnetic moment. This indicated that ^{31}Mg is inside the island. Gade et al. [16, 17] were able to use the two-proton knock-out reactions ^{9}Be(^{38}Si, ^{35,36}Mg + γ) to place ^{35}Mg and ^{36}Mg inside the island to the east (high-N).

Finally, the northern border has been established by work from Lică et al. [18] and Han et al. [19] using β-γ decay measurements; and Xu et al. [20] using β-NMR/β-detected nuclear quadrupole resonance techniques. These studies show that the aluminium isotopes form the boundary of the island, but are not inside it themselves.

The changes in single-particle structure are well illustrated by the magnesium isotopes, as shown in Fig. 1.4 [22]. In ^{32}Mg, the ground state is deformed with an excited spherical configuration; the reverse occurs in ^{30}Mg. This inversion lends its name to this region of the nuclear chart.

1.3 The Structure of ^{29}Mg

To fully understand the phenomena described in this chapter, it is essential to map the evolution of shell structure in the nuclei in this neutron-rich region of the nuclear chart. This work presents an investigation into the structure of ^{29}Mg, which has 17 neutrons. The $N = 17$ isotones are one nucleon outside of the $N = 16$ shell gap. Obtaining systematic information on the population of their orbitals and single-particle energies would therefore help to clarify the evolution of this shell. Between ^{30}Si and ^{28}Mg, the protons would naïvely be removed from the $\pi d_{5/2}$ orbital. Probing the single-particle properties in ^{29}Mg, one neutron outside $N = 16$, allows quantification of the monopole shifts due to the pn-interaction between the proton $d_{5/2}$ and neutron $d_{3/2}$ and fp-orbitals.

Furthermore, the magnesium isotopic chain exhibits the most dramatic transition to the IoI, which can be seen in Fig. 1.3, where the ground states change abruptly from a sd-shell configuration to a pf-shell configuration. Of particular interest is the evolution of intruder configurations in this region. Magnesium-29 acts as a benchmark for these investigations as the last odd-A system outside of the $N = 20$ IoI. Identifying single-particle strength in ^{29}Mg, compared with other measurements within the island, will help reveal the evolution of single-particle structure across this boundary into the island. This change has been characterised using many different reaction probes, but few are sensitive to the underlying single-particle properties.

The chosen method for measuring these properties is the single-neutron transfer reaction $d(^{28}\text{Mg}, p)^{29}\text{Mg}$. The theory of transfer reactions will be discussed further in Chap. 3, describing how single-particle strength can be extracted from measured cross sections.

As is necessary with short-lived nuclei, this reaction was carried out in inverse kinematics using a novel experimental technique that utilises the kinematic advantages of a solenoidal spectrometer, detailed in Chap. 4.

This reaction is well suited to the study of single-particle properties of nuclei. It allows a measurement of the excitation energy of populated states, assignment of relevant angular momentum quantum numbers, and extraction of spectroscopic factors. The centroids of single-particle strength can be determined and compared to data from neighbouring nuclei to assess changes in these centroids, and compared to modern shell-model calculations to assess their description of this region. The methods used in analysing the data are described in Chap. 5 and interpretation and conclusions drawn in Chap. 6.

References

1. Geiger H, Marsden E (1909) On a diffuse reflection of the a-particles. Proc R Soc Lond A 82(557):495–500
2. Geiger H (1910) The scattering of a-particles by matter. Proc R Soc Lond A 83(565):492–504
3. Stroberg SR, Holt JD, Schwenk A, Simonis J (2021) Ab initio limits of atomic nuclei. Phys Rev Lett 126:022501
4. Hoffman CR, Baumann T, Bazin D, Brown J, Christian G, Denby DH, Deyoung PA, Finck JE, Frank N, Hinnefeld J et al (2009) Evidence for a doubly magic ^{24}O. Phys Lett B 672(1):17–21
5. Kanungo R, Nociforo C, Prochazka A, Aumann T, Boutin D, Cortinagil D, Davids B, Diakaki M, Farinon F, Geissel H et al (2009) One-neutron removal measurement reveals ^{24}O as a new doubly magic nucleus. Phys Rev Lett 102:152501
6. Leistenschneider E, Dunling E, Bollen G, Brown BA, Dilling J, Hamaker A, Holt JD, Jacobs A, Kwiatkowski AA, Miyagi T et al (2021) Precision mass measurements of neutron-rich scandium isotopes refine the evolution of N = 32 and N = 34 shell closures. Phys Rev Lett 126:042501
7. Otsuka T, Utsuno Y, Fujimoto R, Brown BA, Honma M, Mizusaki T (2002) Frontiers and challenges of the nuclear shell model. Eur Phys J A 13:69–74
8. Thibault C, Klapisch R, Rigaud C, Poskanzer AM, PrieelsR Lessard L, Reisdorf W (1975) Direct measurement of the masses of ^{11}Li and $^{26-32}$Na with an on-line mass spectrometer. Phys Rev C 12:644–657

9. Warburton EK, Becker JA, Brown BA (1990) Mass systematics for A = 29–44 nuclei: the deformed A 32 region. Phys Rev C 41:1147–1166
10. Brown BA (2010) Islands of insight in the nuclear chart. Physics 3:104
11. Butler PA, Cederkall J, Reiter P (2017) Nuclear-structure studies of exotic nuclei with MINIBALL. J Phys G: Nucl Part Phys 44(4):044012
12. Kobayashi N, Nakamura T, Kondo Y, Tostevin JA, Aoi N, Baba H, Barthelemy R, Famiano MA, Fukuda N, Inabe N et al (2016) One-neutron removal from ^{29}Ne: defining the lower limits of the island of inversion. Phys Rev C 93:014613
13. Gaudefroy L, Mittig W, Orr NA, Varet S, Chartier M, Roussel-Chomaz P, Ebran JP, Fernández-Domínguez B, Frémont G, Gangnant P et al (2012) Direct mass measurements of ^{19}B, ^{22}C, ^{29}F, ^{31}Ne, ^{34}Na and other light exotic nuclei. Phys Rev Lett 109:202503
14. Revel A, Sorlin O, Marqués FM, Kondo Y, Kahlbow J, Nakamura T, Orr NA, Nowacki F, Tostevin JA, Yuan CX et al (2020) Extending the southern shore of the island of inversion to ^{28}F. Phys Rev Lett 124:152502
15. Neyens G, Kowalska M, Yordanov D, Blaum K, Himpe P, Lievens P, Mallion S, Neugart R, Vermeulen N, Utsuno Y et al (2005) Measurement of the spin and magnetic moment of ^{31}Mg: evidence for a strongly deformed intruder ground state. Phys Rev Lett 94:022501
16. Gade A, Adrich P, Bazin D, Bowen MD, Brown BA, Campbell CM, Cook JM, Ettenauer S, Glasmacher T, Kemper KW et al (2007) Spectroscopy of ^{36}Mg: interplay of normal and intruder configurations at the neutron-rich boundary of the "island of inversion". Phys Rev Lett 99:072502
17. Gade A, Bazin D, Brown BA, Campbell CM, Cook JM, Ettenauer S, Glasmacher T, Kemper KW, McDaniel S, Obertelli A et al (2011) In-beam g-ray spectroscopy of ^{35}Mg and ^{33}Na. Phys Rev C 83:044305
18. Lică R, Rotaru F, Borge MJG, Grévy S, Negoiţă F, Poves A, Sorlin O, Andreyev AN, Borcea R, Costache C et al (2017) Identification of the crossing point at N = 21 between normal and intruder configurations. Phys Rev C 95:021301
19. Han R, Li XQ, Jiang WG, Li ZH, Hua H, Zhang SQ, Yuan CX, Jiang DX, Ye YL, Li J et al (2017) Northern boundary of the "island of inversion" and triaxiality in ^{34}Si. Phys Lett B 772:529–533
20. Xu ZY, Heylen H, Asahi K, Boulay F, Daugas JM, de Groote RP, Gins W, Kamalou O, Koszorús Á, Lykiardopoulou M et al (2018) Nuclear moments of the low-lying isomeric 1+ state of ^{34}Al: investigation on the neutron 1p1h excitation across N = 20 in the island of inversion. Phys Lett B 782:619–626
21. Kinsey RR, Dunford CL, Tuli JK, Burrows TW (1996) The NUDAT/PCNUDAT Program for Nuclear Data. Paper submitted to the 9th International Symposium of Capture Gamma-Ray Spectroscopy and Related Topics, Budapest, Hungary, Oct 1996. Data extracted from the NUDAT database, version 2.8. Accessed 4 May 2020. https://www.nndc.bnl.gov/nudat2/
22. Wimmer K, Kröll T, Krücken R, Bildstein V, Gernhäuser R, Bastin B, Bree N, Diriken J, Van Duppen P, Huyse M et al (2010) Discovery of the shape coexisting 0+ state in ^{32}Mg by a two neutron transfer reaction. Phys Rev Lett 105:252501

Chapter 2
The Shell Model

This chapter summarises the primary features of the nuclear shell model, its mathematical framework, and the effect of including residual interactions. It also describes the nucleon-nucleon interaction with emphasis on effects that are relevant to this work.

2.1 The Nuclear Shell Model

Similar to the descriptions of electrons in atomic orbitals, nuclei can be described in terms of occupying nuclear orbitals. For electrons, the only force necessary to describe these orbitals is the Coulomb force, which is mediated by the uncharged photon. Conversely for nucleons, interactions are dominated by the strong interaction which are mediated by the colour-charged gluon. The requirement to model both of these forces for nucleons leads to a difficult and complex problem.

2.1.1 The Nuclear Force

The features of the nuclear force can be inferred from experimental observables [1, 2]:

- The nuclear force must be attractive in order to bind nucleons and overcome the Coulomb force of protons. It must also be short range—this can be understood from Fig. 2.1 which shows the binding energy per nucleon for stable nuclides. This shows a sharp increase with the number of nucleons, until it begins to saturate around $A = 10$, and has a maximum value at ^{62}Ni. This suggests that each nucleon only interacts with a limited number of its neighbours, otherwise the binding energy

The original version of this chapter was revised: The symbol "$\hbar$" has been updated in this chapter. The correction to this chapter can be found at https://doi.org/10.1007/978-3-031-19119-0_7.

P. T. MacGregor, *Single-Particle Structure of ^{29}Mg on the Approach to the N=20 Island of Inversion*, Springer Theses, https://doi.org/10.1007/978-3-031-19119-0_2

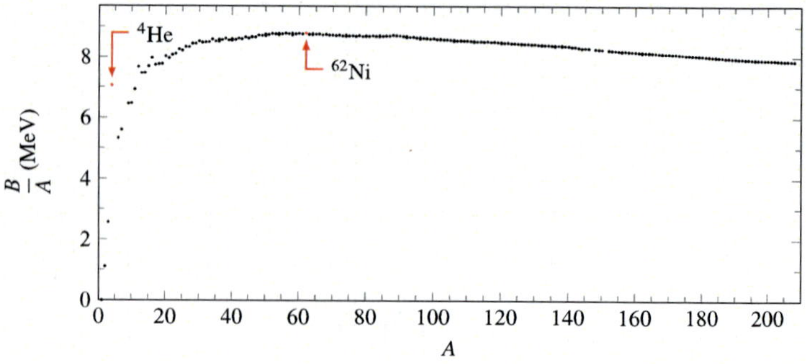

Fig. 2.1 The binding energy per nucleon for stable nuclides. Data from Ref. [3]

would increase further with increasing A. The nuclear force is therefore limited in range to scales of around the size of a nucleon (approximately 1 fm).

- The nucleon-nucleon interaction becomes repulsive at short distances. This is evidenced by singlet s-wave phase shifts in nucleon-nucleon scattering. These phase shifts are an indication of the attractive or repulsive nature of the nuclear potential. For free particles with sufficiently high energy to approach the nuclear core, the nuclear potential becomes repulsive, indicated by a change in sign of these phase shifts.
- The density of the nuclear interior is approximately constant. Electron scattering measurements indicate that the charge distribution of protons in nuclei approximately follow a Fermi distribution, remaining constant for the majority of the interior before falling to zero at large radii. This is consistent with the saturation property and the repulsive core of the nucleon-nucleon interaction mentioned above. Overall, this suggests that the nuclear volume is proportional to A, leading to the familiar relationship to the nuclear radius, $R \propto A^{\frac{1}{3}}$.
- The nuclear force between two protons is similar to that between two neutrons when corrected for Coulomb effects. This can be seen in mirror nuclei, which are pairs of nuclei which have exchanged numbers of protons and neutrons (e.g. $^{29}_{12}Mg_{17}$ and $^{29}_{17}Cl_{12}$). For these, the binding energies are equal when corrected for Coulomb effects, and they have similar energy levels. This is known as charge symmetry.
- The nucleon-nucleon interaction is independent[1] of the nucleons involved in the interaction. This is known as charge independence, and is evidenced by the presence of the same excited states across isobaric chains when corrected for Coulomb effects. This is also the grounds for isospin symmetry, where protons and neutrons can be considered as two states of a nucleon, characterised by the third component of their isospin which is described in quantum mechanics in a similar way to angular momentum for spin-$\frac{1}{2}$ particles.
- The nuclear force is spin dependent, and two nucleons with parallel spins interact more strongly than with antiparallel spins. This is evidenced in deuterium, where

[1] There is a slight difference in the strengths of the nucleon-nucleon interactions, which is evidenced by a small difference between the *np* and *nn*/*pp* scattering lengths. The source of this discrepancy will be discussed in Sect. 2.3.

the only bound state occurs for $S = 1$, when the spins are parallel. The $S = 0$ states are unbound.

- The nuclear force is not spherically symmetric, and contains a non-central component; it has the form $V(\mathbf{r})$ and not $V(r)$. This is again evidenced by deuterium, which has a non-zero quadrupole moment due to a small admixture of $\ell = 2$ into its otherwise spherically symmetric $\ell = 0$ intrinsic ground state.
- The nuclear force depends on the relative angular momentum and spin of the interacting particles. Polarised beams, where the spin of the beam is aligned in a certain direction, scatter differently depending on the orientation of the spin of the beam particles.

In principle, a nuclear force could be constructed that accounts for all of these effects, placed into the Schrödinger equation, and solved. However, solving in this way for multiple nucleons, and including two- and three-body forces, while possible for lighter nuclei (up to around ^{40}Ca), is still beyond the current available computational power for heavier nuclei. It is simpler to approximate this with a mean-field potential that represents the potential created by the nucleons in the nucleus.

2.1.2 Mean-Field Potentials

A two-body Hamiltonian, $\hat{H}$, acting on the wave function of a nucleus, $|\Psi_i\rangle$, can be written as

$$\hat{H}\,|\Psi_i\rangle = \left[-\frac{\hbar^2}{2m}\sum_{j=1}^{A}\nabla_j^2 + \frac{1}{2}\sum_{j=1}^{A}\sum_{k\neq j}\hat{V}_{jk}\right]|\Psi_i\rangle = E_i\,|\Psi_i\rangle\,, \tag{2.1}$$

where $\hat{V}_{jk}$ is the two-body potential operator between nucleons j and k. This Hamiltonian can be split into two parts by the addition and subtraction of a mean-field potential, so that Eq. 2.1 becomes

$$\hat{H}\,|\Psi_i\rangle = \Bigg[\underbrace{\sum_{j=1}^{A}\left(-\frac{\hbar^2}{2m}\nabla_j^2 + U(r_j)\right)}_{\hat{H}^{(0)}} + \underbrace{\left(\frac{1}{2}\sum_{j=1}^{A}\sum_{k\neq j}\hat{V}_{jk} - \sum_{j=1}^{A}U(r_j)\right)}_{\hat{H}^{(1)}}\Bigg]|\Psi_i\rangle\,. \tag{2.2}$$

This mean-field potential is chosen to minimise the residual Hamiltonian, $\hat{H}^{(1)}$, by describing the sum of the two-body potential as best as possible. The independent particle model (IPM) neglects $\hat{H}^{(1)}$, and allows the shell-model Hamiltonian, $\hat{H}^{(0)}$, to be separated into A single-particle Hamiltonians,

$$\hat{H}^{(0)} = \sum_{j=1}^{A}\hat{h}_j\,. \tag{2.3}$$

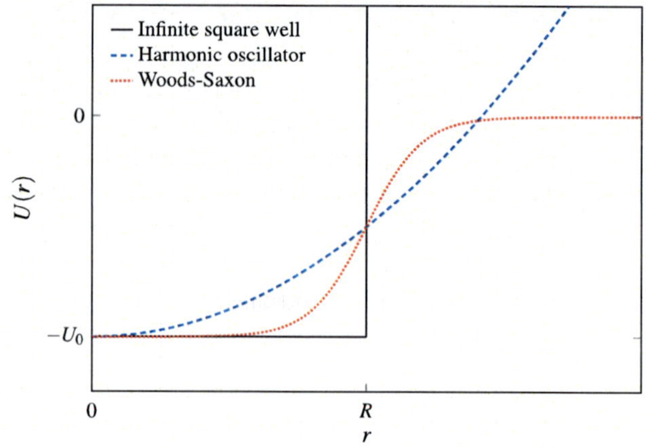

Fig. 2.2 Examples of mean-field potentials plotted as a function of r. Here, R represents the nuclear radius, and U_0 represents the well depth. The forms of these potentials are in Eq. 2.4

Table 2.1 Magic numbers derived from different potentials compared to experimental values [4, 5]

Name	Magic numbers
Infinite square well	**2, 8, 20**, 34, 58, 92, 138
Harmonic oscillator	**2, 8, 20**, 40, 70, 112
Woods-Saxon	**2, 8, 20**, 40, 58, 92, 112
Experiment[a]	2, 8, 20, 28, 50, 82, 126

[a] Along the line of β-stability

This system can then be solved with a suitable choice of the mean-field potential. A number of simple choices for these potentials are shown in Fig. 2.2.

The form of these potentials are:

$$\text{Infinite-square well: } U(r) = \begin{cases} -U_0, & 0 \leqslant r < R, \\ \infty, & \text{otherwise} \end{cases}, \tag{2.4a}$$

$$\text{Harmonic oscillator: } U(r) = \frac{1}{2} m\omega^2 r^2 - U_0, \tag{2.4b}$$

$$\text{Woods-Saxon: } U(r) = \frac{-U_0}{1 + \exp\left(\frac{r-R}{a}\right)}, \tag{2.4c}$$

where U_0 is the well depth, R is the radius of the well, a is the surface diffuseness, m is the mass of the nucleon, and ω is the frequency that must be fitted to the particular nucleus.

Each of these simple potentials has benefits and drawbacks. The infinite square well is the simplest, but does not have realistic properties due to its discontinuity at R which forces all states to be bound. The harmonic oscillator is able to be solved analytically, and, while continuous, has a potential that increases indefinitely so is limited to modelling bound states. The Woods-Saxon, while continuous and tending to 0 for large r, requires numerical solutions to be found.

As shown in Table 2.1, all these potentials fail to reproduce the observed locations of all the "magic numbers". These correspond to the location of shell gaps between nuclear orbitals. An additional term is required in the mean-field potential in order to reproduce these numbers successfully.

2.1.3 *Spin-Orbit Coupling*

Nuclei with a magic number of protons and neutrons were discovered to be particularly stable, but the origin of their stability was unknown for some time. This mystery was resolved by Mayer and Jensen, who discovered that the addition of a spin-orbit term to the potential could account for the magic numbers [6–8], for which they were awarded the Nobel prize in physics in 1963.

The spin-orbit term couples the orbital angular momentum, **L**, with the spin of the nucleus, **S**, and can be written as

$$U_{so}(r) = -U_{\ell s}\frac{1}{r}\frac{\partial U}{\partial r}\mathbf{L}\cdot\mathbf{S}. \tag{2.5}$$

This contains the differential of the form of the nuclear potential, though not necessarily with the same parameters used for the nuclear potential, such as radius and surface diffuseness. The strength of this effect is dictated by $U_{so}(r)$, and it makes orbitals with $j = \ell + \frac{1}{2}$ more energetically favourable than orbits with $j = \ell - \frac{1}{2}$. This effect is strong at the surface of the nucleus, where the potential is changing rapidly.

This addition to the shell model is shown in Fig. 2.3, which shows the evolution of orbitals from harmonic oscillator, to Woods-Saxon, to Woods-Saxon with a spin-orbit term.

2.2 Residual Interactions and Configuration Mixing

While the discussion thus far has focused on the effects of the shell-model Hamiltonian, the residual Hamiltonian cannot be neglected indefinitely, and a more rigorous approach is required.

The wave function of a nucleus in state i, $|\Psi_i\rangle$, can be written using single-particle configurations as a basis,

$$|\Psi_i\rangle = \sum_{c=1}^{N} a_{ci}\,|\Phi_c\rangle\,, \tag{2.6}$$

where $|\Phi_c\rangle$ represents a particular single-particle configuration. This is the eigenbasis of the IPM Hamiltonian,

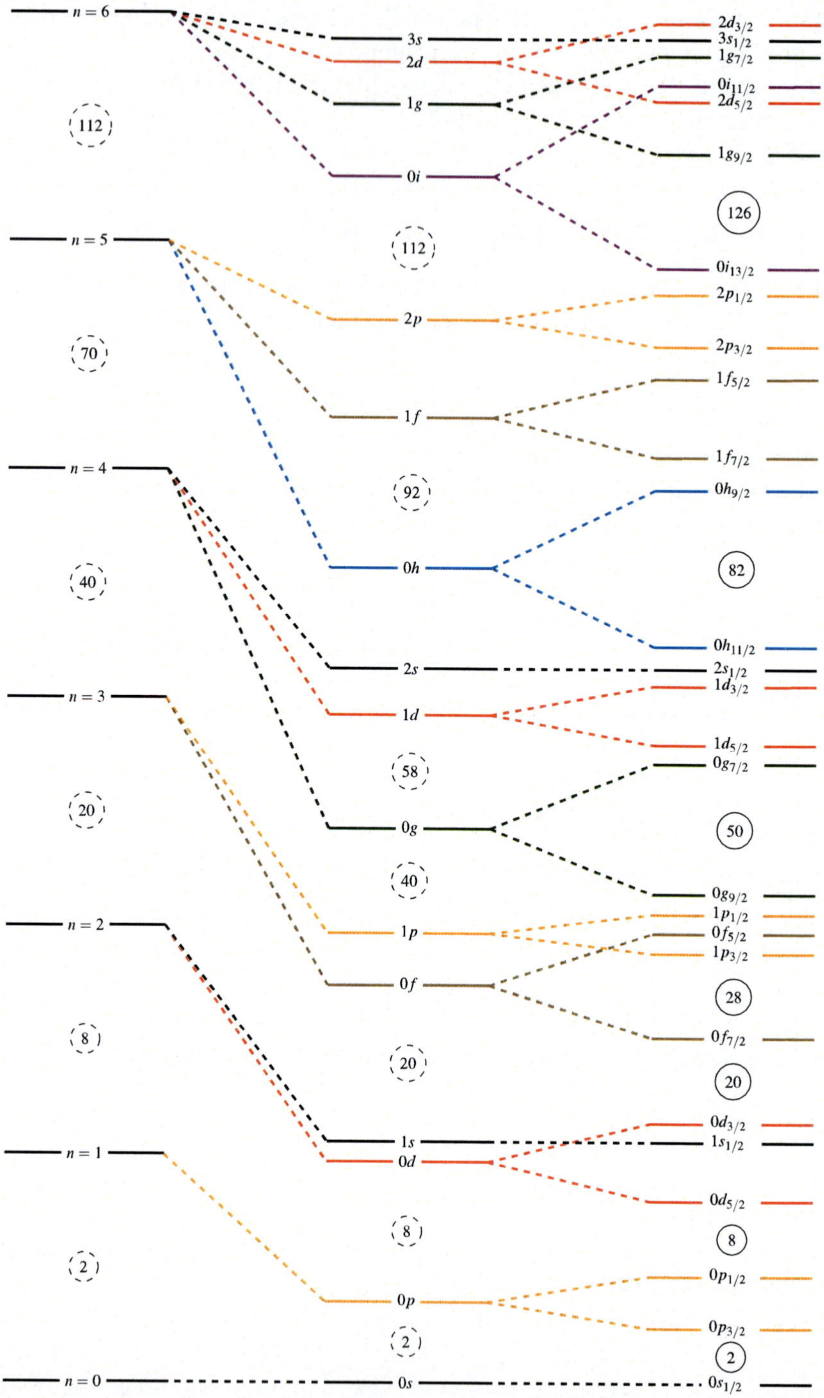

Fig. 2.3 A diagram of the orbitals in the shell model with the respective magic numbers. The left hand side shows the harmonic oscillator orbitals, the middle shows the Woods-Saxon orbitals, and the right hand side shows the splitting that occurs with the addition of a spin-orbit term

$$\hat{H}^{(0)} |\Phi_c\rangle = \sum_{j=1}^{A} \hat{h}_j |\Phi_c\rangle = \sum_{j=1}^{A} \varepsilon_{c,j} |\Phi_c\rangle = \varepsilon_c |\Phi_c\rangle , \tag{2.7}$$

with ε_c representing the energy of a particular configuration. The contribution of each configuration to the total admixture that forms the wave function is represented by a_{ci}. The probability of the nucleus in admixture i being in single-particle configuration c is therefore $a^*_{ci}a_{ci}$. The total number of configurations that form the particular wave function is represented by N. Each of these single-particle configurations can be represented by

$$|\Phi_c\rangle = \alpha^{\dagger}_{A,c} \cdot \alpha^{\dagger}_{A-1,c} \cdot \ldots \cdot \alpha^{\dagger}_{1,c} |0\rangle , \tag{2.8}$$

where $\alpha^{\dagger}_{n,c}$ represents a creation operator for putting nucleon n into a particular single-particle orbital to form configuration $|\Phi_c\rangle$, and $|0\rangle$ is a vacuum state [9]. These operators can have particular commutator relations chosen such that $|\Phi_c\rangle$ is fully antisymmetric.[2] These configurations form an orthonormal basis:

$$\langle \Phi_c | \Phi_d \rangle = \delta_{cd}. \tag{2.9}$$

The expectation value of the energy from the total wave function, $E_i^{(0)}$, can be found by finding the representation of the Hamiltonian operator in this space,

$$\left\langle \Psi_i \left| \hat{H}^{(0)} \right| \Psi_i \right\rangle = \sum_{c=1}^{N} \sum_{d=1}^{N} a^*_{di} a_{ci} \left\langle \Phi_d \left| \hat{H}^{(0)} \right| \Phi_c \right\rangle = \sum_{c=1}^{N} a^*_{ci} a_{ci} \varepsilon_c = E_i^{(0)}, \tag{2.10}$$

which is simply the sum of the product of the single-particle energy, ε_c, weighted by the probability $a^*_{ci}a_{ci}$. For the IPM, where the total wavefunction is a single IPM configuration, $|\Psi_i\rangle = |\Phi_c\rangle$, then the energy $E_i^{(0)} = \varepsilon_c$.

The residual interaction introduces mixing terms to the Hamiltonian. Projecting with a specific single-particle configuration on the full Hamiltonian yields

$$\left\langle \Phi_d \left| \hat{H} \right| \Psi_i \right\rangle = \sum_{c=1}^{N} a_{ci} \underbrace{\left\langle \Phi_d \left| \hat{H} \right| \Phi_c \right\rangle}_{H_{dc}} = \sum_{c=1}^{N} a_{ci} \left[\varepsilon_c \delta_{dc} + \underbrace{\left\langle \Phi_d \left| \hat{H}^{(1)} \right| \Phi_c \right\rangle}_{M_{dc}} \right] = E_i a_{di}, \tag{2.11}$$

where M_{dc} has been used to label the mixing term. This can be more clearly written as a matrix equation:

$$\begin{pmatrix} \varepsilon_1 + M_{11} & M_{12} & \ldots & M_{1N} \\ M_{21} & \varepsilon_2 + M_{22} & \ldots & M_{2N} \\ \vdots & \vdots & \ddots & \vdots \\ M_{N1} & M_{N2} & \ldots & \varepsilon_N + M_{NN} \end{pmatrix} \begin{pmatrix} a_{1i} \\ a_{2i} \\ \vdots \\ a_{Ni} \end{pmatrix} = E_i \begin{pmatrix} a_{1i} \\ a_{2i} \\ \vdots \\ a_{Ni} \end{pmatrix}. \tag{2.12}$$

[2] This is equivalent to forming a Slater determinant with the wave function, forcing the wave function to be antisymmetric, and satisfying the Pauli exclusion principle.

Now, E_i is an eigenvalue of the full Hamiltonian, and can take multiple values upon measurement of the system. The structure of the Hamiltonian shows how the single-particle orbitals mix. If there is no interaction ($M_{dc} = 0$), this reverts to the eigenbasis with values of ε_c on the diagonals. It is worth noting that $M_{dc} = M_{cd}$ because $\hat{H}^{(1)}$ is Hermitian, so the mixing is symmetric. Finally, the diagonal terms in this matrix are the unmixed energy levels, so this can be labelled

$$\varepsilon_a + M_{aa} \equiv E_a \quad \forall a \in \{1, \ldots, N\}. \tag{2.13}$$

To distinguish the unmixed energies from eigenvalues of the Hamiltonian (labelled E_i in Eq. 2.12), the eigenvalues of the Hamiltonian will be labelled with Roman numerals, and the unmixed energies with Arabic numerals.

To obtain these eigenvalues, the Hamiltonian matrix must be diagonalised, so the determinant

$$\begin{vmatrix} E_1 - E_i & M_{12} & \ldots & M_{1N} \\ M_{21} & E_2 - E_i & \ldots & M_{2N} \\ \vdots & \vdots & \ddots & \vdots \\ M_{N1} & M_{N2} & \ldots & E_N - E_i \end{vmatrix} = 0 \tag{2.14}$$

where $i \in \{\text{I, II}, \ldots, \text{N}\}$.

For nuclei, there are a great number of configurations of protons and neutrons that will mix to form single-particle states. It is, therefore, necessary to truncate the model space so that only the most significant configurations for the specific excitation energy range in question are considered.

Often, the nucleus is treated as an inert core with valence nucleons, and these are limited to exist in a finite number of shells outside the core. This is normally sufficient, particularly near closed shells, as it becomes increasingly unrealistic that nucleons will exist in higher single-particle orbitals for a nuclear state at low excitation.

A two-state system is excellent for illustrating the effects of residual interactions. A diagram of a two-state system is shown in Fig. 2.4.

For this system, Eq. 2.14 becomes

$$\begin{vmatrix} E_1 - E_i & M \\ M & E_2 - E_i \end{vmatrix} = 0 \tag{2.15}$$

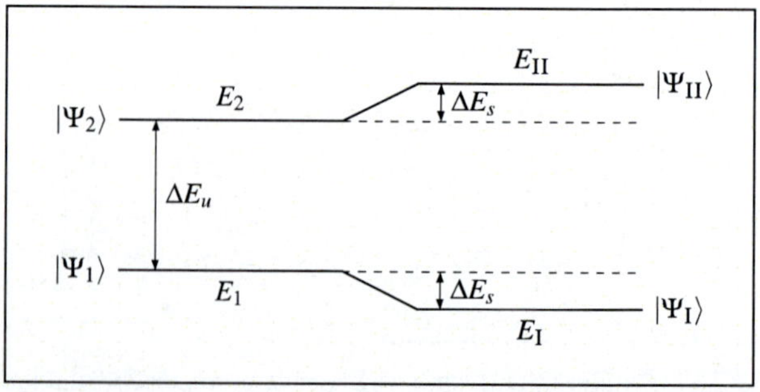

Fig. 2.4 An illustration of residual interactions in a two-state system. Adapted from Ref. [1]

where $M_{12} = M_{21} = M$ for simplicity, and $i \in \{\text{I}, \text{II}\}$. The new energies are therefore

$$E_{\text{I,II}} = \frac{1}{2}(E_1 + E_2) \pm \frac{1}{2}\sqrt{(E_2 - E_1)^2 + 4M^2}. \tag{2.16}$$

Writing this in terms of the initial energy splitting, $\Delta E_u = E_2 - E_1$ yields

$$E_{\text{I,II}} = \frac{1}{2}(E_1 + E_2) \pm \frac{\Delta E_u}{2}\sqrt{1 + \left(\frac{2M}{\Delta E_u}\right)^2}. \tag{2.17}$$

This shows that the net effect of the mixing is repulsive, as $E_{\text{I}} < E_1$ and $E_{\text{II}} > E_2$. This can be used to write the energy splitting ΔE_s,

$$\Delta E_s = |E_{\text{II}} - E_2| = |E_1 - E_{\text{I}}| = \frac{\Delta E_u}{2}\left[\sqrt{1 + \left(\frac{2M}{\Delta E_u}\right)^2} - 1\right]. \tag{2.18}$$

The minimum separation between the new states is therefore M, which occurs when the two energy states are degenerate. Mixed states cannot be closer than this.

This simple picture is complicated when additional states are allowed to mix. As more nucleons interact, with different strengths of the coupling between their respective orbitals, the simple single-particle strength becomes fragmented across a number of different orbitals. Residual interactions such as these are very important when examining the evolution of single-particle structure.

2.3 The Nucleon-Nucleon Interaction and the Evolution of Shell Structure

Residual interactions are required to include the full effect of the nucleon-nucleon interaction term in the Hamiltonian, $\hat{V}_{jk}$. It cannot be neglected in the understanding of nuclear structure and how this shell structure evolves. The components of the nucleon-nucleon interaction can aid in the understanding of the evolution of shell structure, and, in particular, the evolution in neutron-rich nuclei. The nature of the interaction and its phenomenological construction will be discussed here, as well as its effects on shell evolution.

2.3.1 The Monopole Shift

A useful way to characterise the evolution of shell structure is by examining the monopole shift across isobaric and isotopic chains. The monopole shift is the angle averaged global effect of an interaction on a single-particle energy. It is so named because, coincidentally, it is the leading term in a multipole expansion of the Hamiltonian that has been decomposed with different possible couplings of angular momentum (for more details see, for example, Ref. [10]). This can be used to calculate the effective single-particle energy (ESPE), which represents the mean effects of a nucleus on a nucleon [11]. The monopole Hamiltonian can be written as:

$$V^T_{j,j'} = \frac{\sum_J (2J+1)\langle jj'|\hat{V}|jj'\rangle_{JT}}{\sum_J (2J+1)}, \tag{2.19}$$

where j and j' are the angular momenta of two nucleons in different orbits, $\hat{V}$ is the two-body interaction operator between these two nucleons, and $\langle jj'|\hat{V}|jj'\rangle_{JT}$ is the matrix element of the state where two nucleons are coupled to angular momentum J and isospin T in a way that satisfies antisymmetrisation [12]. This is averaged over J to access the mean effect of this shift.

The shift in the single-particle energy for neutrons in orbit j' with the addition of a proton into orbit j is

$$\Delta\varepsilon_n(j) = \frac{1}{2}\left[V^{T=0}_{j,j'} + V^{T=1}_{j,j'}\right] n_p(j'), \tag{2.20}$$

where $n_p(j')$ is the expectation value of the number of protons in orbit j'. The converse is also true for the shift in energy of protons due to the addition of a neutron [12].

2.3.2 *Central Forces and the Spin-Orbit Interaction*

The IPM models the nucleon-nucleon interaction with a mean-field central potential. This is a good approximation as central forces are the major contributor to the nucleon-nucleon interaction. Assuming that the interaction depends only on the two-nucleon separation, **r**, the relative linear momentum, **p**, the nucleon spin $\underset{\sim}{\sigma}^{(i)}$, and the nucleon isospin, $\vec{\tau}_i$, then symmetry arguments imply that the central potential between two nucleons has the general form

$$U_c(r) = \underbrace{\left[V_c(r) + \tilde{V}_c(r)\vec{\tau}_1\cdot\vec{\tau}_2\right]}_{\text{central}} + \underbrace{\left[V_{cs}(r) + \tilde{V}_{cs}(r)\vec{\tau}_1\cdot\vec{\tau}_2\right]\left(\underset{\sim}{\sigma}^{(1)}\cdot\underset{\sim}{\sigma}^{(2)}\right)}_{\text{central spin dependent}}, \tag{2.21}$$

where a number of symmetries have been respected[3] [13]. The functions V_c, $\tilde{V}_c$, V_{cs}, and $\tilde{V}_{cs}$ are all radial functions that are used to model the nuclear potential, with the property that they tend to 0 as $r \to \infty$.

A similar treatment can be applied to the spin-orbit potential in Eq. 2.5, and has the form

$$U_{so}(\mathbf{r}) = \left[V_{so}(r) + \tilde{V}_{so}(r)\vec{\tau}_1 \cdot \vec{\tau}_2\right] \underbrace{(\mathbf{r} \times \mathbf{p})}_{\mathbf{L}} \cdot \underbrace{\left(\underset{\sim}{\sigma}^{(1)} + \underset{\sim}{\sigma}^{(2)}\right)}_{\underset{\sim}{\sigma} \propto \mathbf{S}}. \tag{2.22}$$

To match the term from Eq. 2.5,

$$V_{so}(r) = -U_{\ell s}\frac{1}{r}\frac{\partial U}{\partial r}. \tag{2.23}$$

As the spin-orbit interaction depends on the derivative of the nuclear potential, the diffuseness of the nuclear surface will impact its strength.

2.3.3 *The Tensor Interaction*

As mentioned in Sect. 2.1.1, the nuclear force was discovered to have a non-central component from the non-zero quadrupole moment in deuterium. The quadrupole moment has the definition

$$Q = \left\langle \Psi_i \left| (3z^2 - r^2) \right| \Psi_i \right\rangle, \tag{2.24}$$

for a system in a state, $|\Psi_i\rangle$, and where z is an arbitrarily chosen quantisation axis. The quadrupole moment is zero for a spherically symmetric system. For the deuteron, $Q = +0.29$ fm^2 experimentally [14], indicating that the deuteron has a charge distribution that is prolate. Combined with experimental data that the total angular momentum is $J = 1$ and the magnetic moment suggesting that the spins are parallel, the only possible conclusion is that the deuteron bound state is an admixture of $^3\mathrm{S}_1$ and $^3\mathrm{D}_1$ states.

Central forces conserve orbital angular momentum, so mixing of $^3\mathrm{S}_1$ and $^3\mathrm{D}_1$ configurations requires a non-central component to the interaction. This non-central or tensor component has the general form

$$U_t(r) = \left[V_t(r) + \tilde{V}_t(r)\vec{\tau}_1 \cdot \vec{\tau}_2\right] S_{12}. \tag{2.25}$$

A similar expression to this can be obtained from first principles using a one-pion exchange potential (OPEP), where the different pions mediate the nuclear force. This has the form [13]:

[3] Namely invariance under reflection and rotation of the coordinate axes, and time reversal; as well as satisfying Hermiticity, charge symmetry and independence, symmetry under two-nucleon exchange, no explicit time-dependence, and conservation of electrical charge [13].

$$V_{\text{OPEP}} \propto \vec{\tau}_1 \cdot \vec{\tau}_2 \left[\underset{\sim}{\sigma}^{(1)} \cdot \underset{\sim}{\sigma}^{(2)} + \left(1 + \frac{3}{\mu r} + \frac{3}{\mu^2 r^2}\right) S_{12} \right] \frac{e^{-\mu r}}{r}, \tag{2.26}$$

where $\mu = \frac{m_\pi c}{\hbar}$. This assumes charge independence, but the neutral pion is slightly lighter than the charged pions which breaks this symmetry. While a more complete picture would require additional mesons and allow for multiple-meson exchanges, the interaction involving only pions is sufficient to match the corresponding phenomenological interaction in Eq. 2.25.

The most important term in Eqs. 2.25 and 2.26 is S_{12}. It gives the tensor characteristic to this force and has the form

$$S_{12} = \frac{3\left(\underset{\sim}{\sigma}^{(1)} \cdot \mathbf{r}\right)\left(\underset{\sim}{\sigma}^{(2)} \cdot \mathbf{r}\right)}{r^2} - \underset{\sim}{\sigma}^{(1)} \cdot \underset{\sim}{\sigma}^{(2)}, \tag{2.27}$$

where $\mathbf{r} = \mathbf{r}_2 - \mathbf{r}_1$. This operator is a scalar, and conserves the nuclear spin, $\mathbf{J}$ and the parity. However, the non-central nature of the potential means that $\mathbf{L}$ and $\mathbf{S}$ are not conserved.

The term S_{12} has different eigenvalues depending on the coupling of the spins. For the singlet $S = 0$ state, $\underset{\sim}{\sigma}^{(1)} = -\underset{\sim}{\sigma}^{(2)}$, so $S_{12} = 0$ irrespective of the orientation of the nucleons. For the $S = 1$ triplet state, S_{12} is non-zero: using the property that $\mathbf{S} = \frac{\hbar}{2}\underset{\sim}{\sigma}$, then for nucleons,

$$\underset{\sim}{\sigma}^{(1)} \cdot \underset{\sim}{\sigma}^{(2)} = \frac{1}{2}\left[\left(\underset{\sim}{\sigma}^{(1)} + \underset{\sim}{\sigma}^{(2)}\right)^2 - \left(\underset{\sim}{\sigma}^{(1)}\right)^2 - \left(\underset{\sim}{\sigma}^{(2)}\right)^2\right] = \begin{cases} +1, & S = 1 \\ -3, & S = 0 \end{cases}, \tag{2.28}$$

and

$$\frac{3\left(\underset{\sim}{\sigma}^{(1)} \cdot \mathbf{r}\right)\left(\underset{\sim}{\sigma}^{(2)} \cdot \mathbf{r}\right)}{r^2} = \begin{cases} +3, & \underset{\sim}{\sigma} \parallel \mathbf{r} \\ 0, & \underset{\sim}{\sigma} \perp \mathbf{r} \end{cases}. \tag{2.29}$$

These can be combined to obtain S_{12}, and this is represented pictorially in Fig. 2.5.

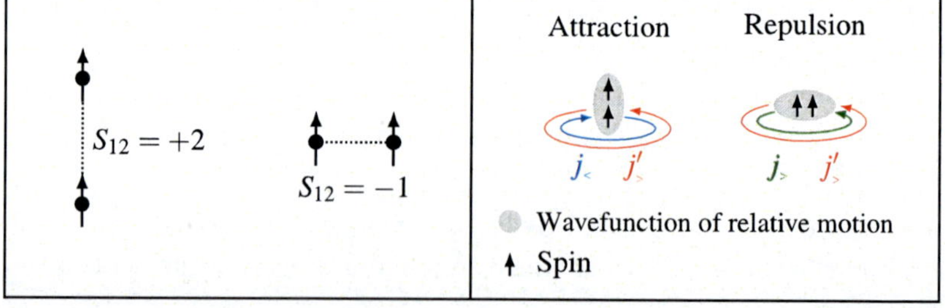

Fig. 2.5 Left: the different combinations of spin orientation and inter-nucleon separation for two nucleons. Adapted from Ref. [15]. Right: the $S = 1$ coupling effect on the tensor interaction. Adapted from Ref. [12]

For $S = 1$, the effect of S_{12} either enhances the force by making it more attractive, or reverses its effect by making it negative. These properties of the tensor interaction are represented pictorially in the right panel of Fig. 2.5 for two nucleons in different orbits. In the first instance, the nucleons have total angular momenta of $j_< = \ell - \frac{1}{2}$ and $j'_> = \ell' + \frac{1}{2}$, and are coupled to an $S = 1$ state. The orbits therefore travel in opposite directions. The relative momentum between them is therefore quite high, and, by the uncertainty principle, the spatial wave function has a narrower distribution. This results in the wave function having a prolate shape which is favoured by the tensor interaction, and the nucleons are attracted to one another. This is the case for the deuteron, where the positive quadrupole moment and the resulting prolate shape of the nucleus arises from the tensor force, producing its only bound state.

Conversely, for two orbits with nucleons travelling in the same direction (i.e. both j's are of the form $j_>$ or $j_<$), such as the second case with nucleons with $j_> = \ell + \frac{1}{2}$ and $j'_> = \ell' + \frac{1}{2}$, the opposite occurs. The wave function has an oblate shape, with a large spatial distribution due to the small relative momentum between the two nucleons. The nucleons are repelled from each other.

The attraction between nucleons is enhanced if the radial components of the wave function have significant overlap, and if the relative momentum between the nucleons is increased. Therefore, this effect is most significant for nucleons with the same or similar ℓ, and for orbits with higher values of ℓ due to the higher relative momentum between the nucleons [12, 16].

An example of the tensor interaction affecting single-particle energies is shown in Fig. 2.6 for proton ESPEs in nuclides of antimony ($Z = 51$) [12]. Antimony has one proton outside the $Z = 50$ shell gap. As neutrons fill the $\nu 0h_{11/2}$ orbital (which is assumed to be uniformly filled as N increases), the spacing between the $\pi 0h_{11/2}$

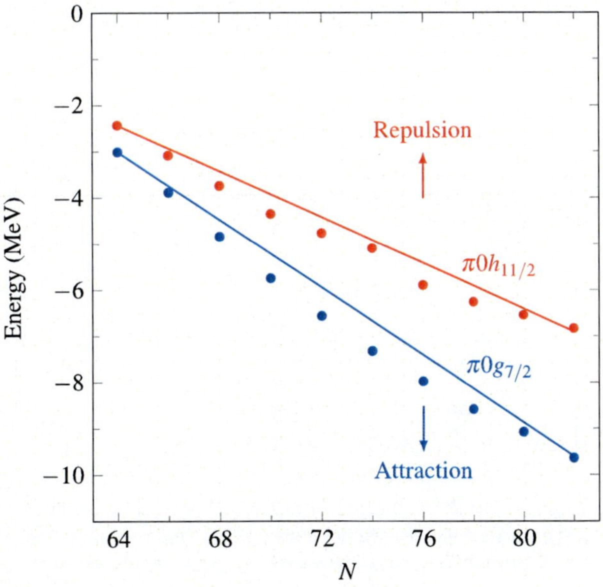

Fig. 2.6 The change in proton ESPEs for isotopes of antimony. The data points follow a similar trend to the theoretical calculations (lines). Adapted from Ref. [12], with data from Ref. [17]

and $\pi 0g_{7/2}$ increases. If the cause were a central interaction, the ESPEs of these orbitals would not diverge. But the tensor interaction can generate both attractive and repulsive effects: the filling $\nu 0h_{11/2}$ orbital ($j = j_> = 5 + \frac{1}{2}$) repels its counterpart proton orbital, and attracts the $\pi 0g_{7/2}$ ($j = j_< = 4 - \frac{1}{2}$), which causes the gap between these orbitals to increase. Calculations with a tensor force (shown in Fig. 2.6) do indeed reproduce the observed trends.

2.3.4 Type-I and Type-II Shell Evolution

Until this point, the shell evolution that has been discussed is Type-I shell evolution [18]. However, it is prudent to discuss and distinguish this from Type-II shell evolution which, while not occurring explicitly in this work, does play a role in the interior of the IoI. These two types of shell evolution are illustrated in Fig. 2.7.

Type-I shell evolution occurs across isotopic or isotonic chains. Assuming that the neutron orbitals fill the lowest single-particle orbit after a closed shell (indicated by the box in Fig. 2.7), then this will affect the energy of the proton orbitals. As the neutron orbital with ($j = j'_>$) is filled, the tensor interaction has an attractive monopole shift for the proton orbital with ($j = j_<$) and a repulsive monopole shift for the proton orbital with ($j = j_>$). This effect is linear, so twice the number of neutrons doubles the magnitude of this reduction in the splitting of orbitals. This effect is, of course, stronger when ℓ' and ℓ have a significant radial overlap [18].

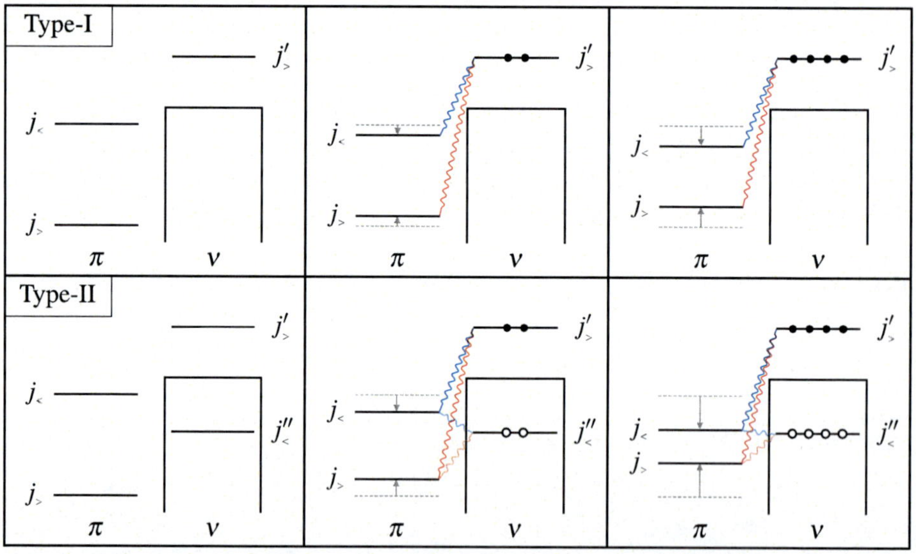

Fig. 2.7 An illustration of Type-I and Type-II shell evolution. The filled circles represent neutrons, the empty circles represent neutron holes, and the wavy lines indicate the action of the tensor force on an orbital. The rectangle indicates the neutron core of the nucleus. Adapted from Ref. [18]

Type-II shell evolution occurs in nuclei where particle-hole excitations are prevalent. Here, as the number of particle-hole excitations increases, the monopole shift acts in two places. Firstly, as in Type-I evolution, the neutrons with $j = j'_>$ attract the protons with $j = j_<$ and repel the protons with $j = j_>$. However, the tensor-force monopole shift from the fully occupied state with $j = j''_<$ is weakened. Therefore, as in Fig. 2.7), the holes with $j = j''_<$ further attract protons with $j = j_<$ and further repel those with $j = j_>$ [18].

This type of shell evolution becomes important for intruder states that occur as the IoI is entered. However, in this work, only Type-I shell evolution is encountered on the approach to the IoI. The Type-I shell evolution that occurs along the $N = 17$ isotones will be discussed further in Chap. 6.

2.3.5 *Summary*

The picture of nuclear shell structure given in this chapter is not static. It has been developed to describe observations in stable nuclei, but deviations from this have been observed in exotic nuclei, away from stability, and in weakly-bound systems. The work presented aims to investigate the evolution of shell structure away from stability, which was summarised in Chap. 1. Shell structure can be probed using transfer reactions, which is the subject of the next chapter.

References

1. Casten RF (1990) Nuclear Structure from a simple perspective, 1st edn. Oxford studies in nuclear physics. Oxford University Press
2. Krane KS (1988) Introductory nuclear physics. Wiley
3. Wang M (王猛), Huang WJ (黄文嘉), Kondev FG, Audi G (欧乔治), Naimi S (2021) The AME 2020 atomic mass evaluation (II). Tables, graphs and references. Chin Phys C 45(3):030003
4. Heyde K (1994) Basic ideas and concepts in nuclear physics. An introductory approach. Fundamental and applied nuclear physics series. IoP Publishing Ltd
5. Bertulani CA (2007) Nuclear physics in a nutshell, vol 2. Princeton University Press
6. Mayer MG (1948) On closed shells in nuclei. Phys Rev 74:235–239
7. Mayer MG (1949) On closed shells in nuclei. II. Phys Rev 75:1969–1970
8. Haxel O, Jensen JHD, Suess HE (1949) On the "magic numbers" in nuclear structure. Phys Rev 75:1766–1766
9. Stroud DG (2002) Notes on occupation number formalism for fermions. Ohio State University. https://www.asc.ohio-state.edu/stroud.2/p882dn1.pdf. Accessed 14 Apr 2021
10. De Maesschalck A (2006) The changing mean field in exotic nuclei: a shell-model point of view. PhD Thesis, UGent. Faculteit Wetenschappen
11. Otsuka T, Utsuno Y, Fujimoto R, Brown BA, Honma M, Mizusaki T (2002) Frontiers and challenges of the nuclear shell model. Eur Phys J A 13:69–74
12. Otsuka T, Suzuki T, Fujimoto R, Grawe H, Akaishi Y (2005) Evolution of nuclear shells due to the tensor force. Phys Rev Lett 95:232502
13. Hodgson PE, Gadioli E, Gadioli-Erba E (1997) Introductory nuclear physics. Oxford University Press

14. Puchalski M, Komasa J, Pachucki K (2020) Hyperfine structure of the first rotational level in H2, D2 and HD molecules and the deuteron quadrupole moment. Phys Rev Lett 125:253001
15. Blatt JM, Weisskopf VF (1952) Theoretical nuclear physics. Wiley. theoreticalnucle00blat_0
16. Kay BP (2007) High-j single-particle strength outside the Z = 50 and N = 82 cores. PhD thesis, The University of Manchester
17. Schiffer JP, Freeman SJ, Caggiano JA, Deibel C, Heinz A, Jiang C-L, Lewis R, Parikh A, Parker PD, Rehm KE et al (2004) Is the nuclear spin-orbit interaction changing with neutron excess? Phys Rev Lett 92:162501
18. Otsuka T, Tsunoda Y (2016) The role of shell evolution in shape coexistence. J Phys G: Nucl Part Phys 43(2):024009

Chapter 3
Transfer Reactions and Reaction Theory

This chapter briefly outlines how transfer reactions can be used as a probe of single-particle structure within the nucleus using reaction theory. This is not intended to be exhaustive, but instead to give a general outline of the theory underpinning the experimental chapters that will follow.

3.1 The Cross Section

An essential tool for studying atomic nuclei is scattering, because it can characterise the nucleus. When a particle is fired at a nucleus and interacts, the properties of the beam and target particles determine which reactions are possible. The yields of different reactions can be characterised by a cross section.

For a given reaction involving a beam of intensity I_0, incident on a target containing N_t particles per unit area, the number of emitted particles (the yield), Y, is given by

$$Y = I_0 N_t \sigma, \tag{3.1}$$

where σ is the total cross section for that particular reaction. This is effectively a measure of the probability that a given reaction can take place. For scattering into a detector of finite size, with a solid angle $\Delta\Omega$, the differential cross section is a more useful quantity. It is generally a function of the polar angle, θ, and the azimuthal angle, ϕ.

The measured differential cross section also must take the absolute efficiency of the detector, ε_δ, and the purity of the beam and target, $\varepsilon_\pi = \varepsilon_\beta \varepsilon_\tau$, into account. All of these variables are illustrated in Fig. 3.1, which shows a schematic illustration of a scattering experiment.

The differential cross section is therefore

The original version of this chapter was revised: The symbol "$\hbar$" has been updated in this chapter. The correction to this chapter can be found at https://doi.org/10.1007/978-3-031-19119-0_7.

P. T. MacGregor, *Single-Particle Structure of ^{29}Mg on the Approach to the N=20 Island of Inversion*, Springer Theses, https://doi.org/10.1007/978-3-031-19119-0_3

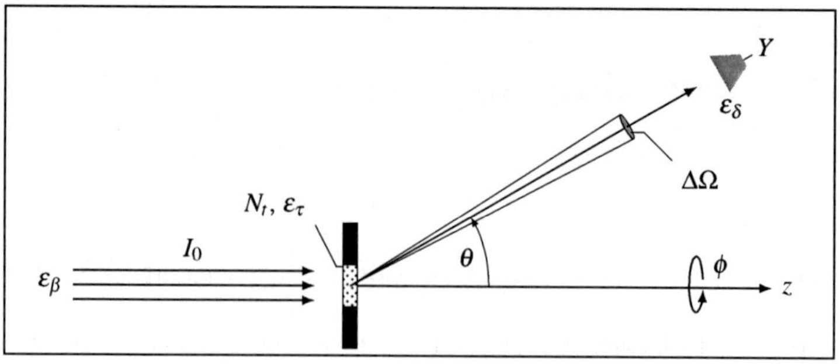

Fig. 3.1 An illustration showing the scattering of a beam with intensity I_0 and purity ε_β interacting with a number of particles per unit area within a target, N_t, with purity ε_τ. The resultant ejectiles are scattered at angles θ and ϕ into a detector of efficiency ε_δ with a solid angle of $\Delta\Omega$. The detector measures a yield, Y

$$\frac{\mathrm{d}\sigma}{\mathrm{d}\Omega}(\theta, \phi) = \frac{Y}{I_0 N_t \cdot \Delta\Omega \cdot \varepsilon_\pi \varepsilon_\delta}. \tag{3.2}$$

A classical example of a reaction is Rutherford scattering, which is the scattering of charged particles from a target nucleus by a purely Coulomb force. As the form of the potential is known, the cross section can be calculated analytically as

$$\frac{\mathrm{d}\sigma}{\mathrm{d}\Omega} = \left(\frac{Z_1 Z_2}{E}\right)^2 \left(\frac{\hbar c \alpha}{4}\right)^2 \frac{1}{\sin^4\left(\frac{\theta}{2}\right)}, \tag{3.3}$$

where Z_1 and Z_2 are the charges of the beam and target in units of e respectively, E is the beam energy, and θ is the angle of scattering [1].

As the scattering potential has a simple form, the cross section can be written analytically. However, nuclear reactions interact primarily through the strong interaction, which has a potential that is difficult to describe in analytical form. Therefore, most nuclear reactions must be modelled to obtain a theoretical cross section.

3.2 Single-Nucleon Transfer Reactions

For this work, the reactions of interest are single-nucleon transfer reactions. This is a direct reaction mechanism, where a nucleon is transferred to or from a nucleus in a single-step process. Therefore, a single degree of freedom in the nucleus is excited, resulting in a large overlap between the wave functions in the entrance and exit channels. This overlap is governed by the matrix element between the initial and final states.

Direct processes differ from compound nuclear mechanisms, where the beam particle fuses with the target and its energy is distributed across the whole nucleus before the ejectile is emitted. When a compound nuclear state is formed before the emission of one or more particles, it is called fusion evaporation. This competes with the single-step direct mechanism. An illustration of this difference is shown in Fig. 3.2 for a deuteron beam and a proton ejectile.

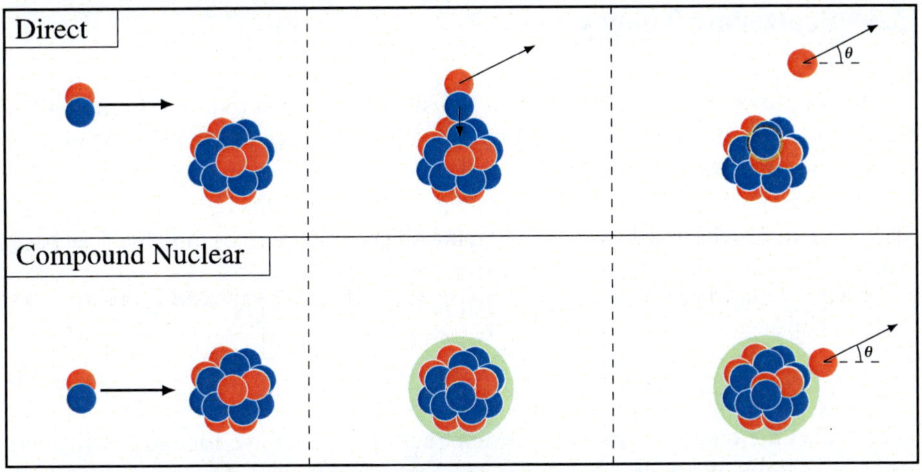

Fig. 3.2 An illustration of a (*d*,*p*) transfer reaction via direct single-step and compound nuclear multi-step processes. Time progresses from left to right

While they can have the same reaction products, the fusion evaporation reaction does not populate single-particle states (where a single degree-of-freedom is excited). Instead a number of degrees of freedom are excited, forming a collective state. While these states are distinct, they often form a continuum corresponding to the many different ways of distributing the energy across the nucleus. This continuum typically happens at higher excitation energies, where the level density is much greater.

In the case of the direct (*d*,*p*) transfer reaction, the neutron enters a single-particle state that is well defined, so the ejected protons come out in a similarly well-defined state, allowing a direct probe of these single-particle states. On the contrary, the corresponding reaction through the compound nucleus state does not have this, as the information of the state that the deuteron enters is lost after the energy and angular momentum is distributed.

While it is impossible to totally reject protons originating from the compound nuclear process in a (*d*,*p*) reaction measurement, the design of the experiment can be adjusted to make the detection of these protons much less probable than those from the direct process. Ejectiles originating from direct processes dominate at beam energies of approximately 10 MeV/u, and have the highest yields at forward angles in the centre-of-momentum (CM) frame. Compound nuclear processes dominate at lower beam energies, with the cross section exhibiting much weaker angle-dependence. Therefore, selecting favourable beam energies and measuring scattered ejectiles at lower CM angles maximises the probability for detecting ejectiles from a direct process for a given reaction.

3.3 Scattering Theory

A quantum-mechanical description of transfer reactions is required to model them accurately. Nuclei are not point particles, and have relative motion to one another as well as internal degrees of freedom. The following will outline how these reactions can be described. The reaction nomenclature A(a,b)B will be used. The entrance channel will be labelled $\alpha = a + \mathrm{A}$, and the exit channel will be labelled $\beta = b + \mathrm{B}$. This will all be done in the CM frame for convenience.

To begin with, the Schrödinger equation for this system is

$$\hat{H}\Psi = E\Psi \tag{3.4}$$

where E is the energy of the total system, and Ψ is the wave function of the entire nuclear system. While this fully describes the system, it is not very useful in this form. It can, however, be decomposed into different parts. The Hamiltonian for the entrance channel can be written as

$$\hat{H} = \hat{H}_\alpha + \hat{T}_\alpha + V_\alpha, \tag{3.5}$$

where $\hat{H}_\alpha = \hat{H}_a + \hat{H}_\mathrm{A}$ describes the Hamiltonian operators for the internal structure of a and A in the CM frame.[1]

The kinetic energy operator of the relative motion between a and A is

$$\hat{T}_\alpha = -\frac{\hbar^2}{2\mu_\alpha}\nabla_\alpha^2, \tag{3.6}$$

where ∇_α acts on $\mathbf{r}_\alpha = \mathbf{r}_a - \mathbf{r}_A$, and μ_α is the reduced mass of the $a + \mathrm{A}$ system. The operator that describes the interaction between all of the nucleons in a and A is V_α. This can be written as the sum over all two-nucleon interactions,

$$V_\alpha = \sum_{i \in a,\, j \in \mathrm{A}} V_{ij}. \tag{3.7}$$

Finally the wave function itself can be decomposed into its constituent parts. For the entrance channel,

$$\Psi_\alpha = \Phi_\alpha \psi_\alpha = \Phi_a \Phi_\mathrm{A} \psi_\alpha \tag{3.8}$$

where Φ_a and Φ_A describe the internal wave functions of a and A respectively. These describe the properties of the constituent nucleons, such as their relative positions, spins, and excitation energies. The wave function, ψ_α, describes the relative motion

---

[1] If $\hat{H}$ were not in the CM frame, an additional component, $\hat{H}_{\mathrm{frame}}$ would also be needed to show that there has been a change of frame. However, this remains constant for the reaction, so can be neglected in any further analysis.

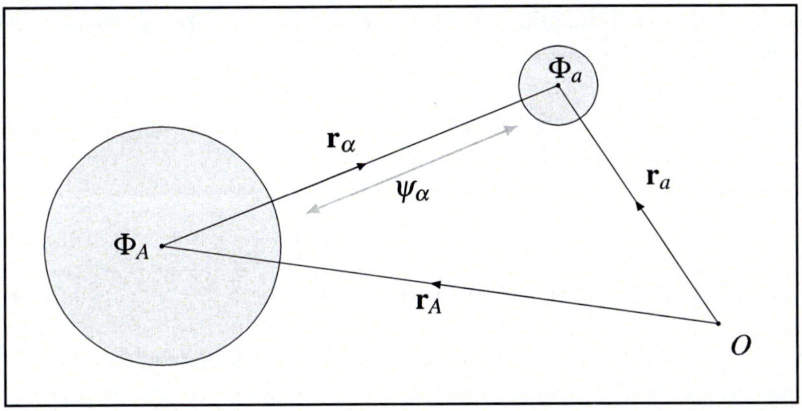

Fig. 3.3 An illustration of the two particles involved when describing a transfer reaction. The wave function ψ_α encodes the relative motion only, whereas the wave functions Φ_a and Φ_A encode the remaining properties of each particle

between a and A. Similar Hamiltonians and wave functions can be constructed for the exit channel, β, as well. This is illustrated in Fig. 3.3.

The goal is to obtain a wave function that describes the relative motion between particles in β in terms of the wave functions in α at a long distance from the interaction site after the transfer reaction. While the following will not derive any analytical solution, projecting Φ_β onto Ψ_α allows the fundamentals of the interaction to be understood.

First, the Schrödinger equation can be rewritten as

$$\left(\hat{H} - \hat{H}_\beta - \hat{T}_\beta\right)\Psi_\alpha = \left(E - \hat{H}_\beta - \hat{T}_\beta\right)\Psi_\alpha = V_\beta \Psi_\alpha. \tag{3.9}$$

This uses the interaction between particles in β to frame the solution that will be obtained. The internal wave function of β can then be projected onto both sides of this equation:

$$\int \mathrm{d}\xi\ \Phi_\beta^* \left(E - \hat{H}_\beta - \hat{T}_\beta\right)\Psi_\alpha = \int \mathrm{d}\xi\ \Phi_\beta^* V_\beta \Psi_\alpha. \tag{3.10}$$

where the internal coordinates of all particles have been labelled with ξ. This means that ψ_α is not a part of the integration, and can be factored out. The operator $\hat{T}_\beta$ will not act on any internal wave function. The Hermitian operator $\hat{H}_\beta$ can also act on Φ_β^* to obtain the internal energy, ε_β. Therefore, the energy term can be simplified further using $E_\beta = E - \varepsilon_\beta$.

It is then possible to define

$$\psi_\beta(\mathbf{r}_\beta) = \left(\int \mathrm{d}\xi\ \Phi_\beta^* \Phi_\alpha\right)\psi_\alpha = \int \mathrm{d}\xi\ \Phi_\beta^* \Psi_\alpha \tag{3.11}$$

where ψ_β is the wave function of relative motion in β, here expressed in terms of the wave function in α. The dependence on $\mathbf{r}_\beta$ has been made explicit here, as this is a spatial wave function. This wave function is maximised when there is little to no rearrangement between nucleons. In the case of elastic scattering, where the entrance and exit channels are identical, then $\psi_\beta = \psi_\alpha$.

Therefore, using Eq. 3.11, Eq. 3.10 can be written as

$$\left(E_\beta - \hat{T}_\beta\right)\psi_\beta = \frac{\hbar^2}{2\mu_\beta}\left(\nabla_\beta^2 + k_\beta^2\right)\psi_\beta = \int \mathrm{d}\xi\ \Phi_\beta^* V_\beta \Psi_\alpha, \tag{3.12}$$

which is a scaled form of the inhomogeneous Helmholtz equation [2]. This can be solved with a Green's function with the boundary condition that it goes to zero as $\mathbf{r} \to \infty$ [3].

The general solution to Eq. 3.12 is

$$\psi_\beta(\mathbf{r}_\beta) = \delta_{\beta\alpha} e^{i\mathbf{k}_\alpha \cdot \mathbf{r}_\alpha} - \frac{2\mu_\beta}{\hbar^2}\int \mathrm{d}^3\mathbf{r}'\ \frac{e^{ik_\beta|\mathbf{r}_\beta - \mathbf{r}'|}}{4\pi\left|\mathbf{r}_\beta - \mathbf{r}'\right|}\int \mathrm{d}\xi\ \Phi_\beta^* V_\beta \Psi_\alpha, \tag{3.13}$$

where $\delta_{\beta\alpha}$ is the Kronecker delta, and only allows an incoming plane wave when the entrance and exit channels are identical. The second term in this can be approximated at large $\mathbf{r}_\beta$, corresponding to a large distance from the interaction site. This is where the exit channel has a well-defined wave function with no external potential acting on it. Using the approximation

$$\lim_{r_\beta \to \infty}\left|\mathbf{r}_\beta - \mathbf{r}'\right| = r_\beta\left(1 - \frac{\mathbf{r}_\beta \cdot \mathbf{r}'}{r_\beta^2}\right), \tag{3.14}$$

then Eq. 3.13 becomes

$$\psi_\beta(\mathbf{r}_\beta) \to \delta_{\beta\alpha} e^{i\mathbf{k}_\alpha \cdot \mathbf{r}_\alpha} - \underbrace{\frac{e^{ik_\beta r_\beta}}{r_\beta}}_{\text{sph. wave}}\ \underbrace{\frac{\mu_\beta}{2\pi\hbar^2}\int \mathrm{d}^3\mathbf{r}'\ e^{-i\mathbf{k}_\beta \cdot \mathbf{r}'}\int \mathrm{d}\xi\ \Phi_\beta^* V_\beta \Psi_\alpha,}_{f_{\beta\alpha}(\theta)} \tag{3.15}$$

where $f_{\beta\alpha}(\theta)$ is the amplitude of the scattered, spherical wave. The angle between $\mathbf{k}_\beta$ and $\mathbf{k}_\alpha$ is θ. This form of the wave function shows that whatever the exit channel is, there will always be some spherical wave (at large distances from the interaction) governed by a particular amplitude.

The term $f_{\beta\alpha}(\theta)$ controls the angular behaviour of the exit channel, and so is related to the cross section. Its relation to the cross section can be calculated using the quantum current. The quantum current in a channel, γ, is defined as

$$\mathbf{j}_\gamma = \frac{\hbar}{2i\mu_\gamma}\left[\psi_\gamma^* \nabla \psi_\gamma - \psi_\gamma \nabla \psi_\gamma^*\right]. \tag{3.16}$$

For the spherical wave in the exit channel, the total current passing through an infinitesimal area $r_\beta^2 \mathrm{d}\Omega$ for large $\mathbf{r}_\beta$ is

$$\frac{\mathbf{j}_\beta \cdot \mathbf{r}_\beta}{r_\beta} r_\beta^2 \,\mathrm{d}\Omega = \frac{\hbar}{2i\mu_\beta}\left[\frac{2ik_\beta}{r_\beta^2}\right]\left|f_{\beta\alpha}(\theta)\right|^2 r_\beta^2 \,\mathrm{d}\Omega = \frac{\hbar k_\beta}{\mu_\beta}\left|f_{\beta\alpha}(\theta)\right|^2 \mathrm{d}\Omega. \tag{3.17}$$

The current passing through a unit area for the incoming plane wave in the entrance channel is

$$\frac{\mathbf{j}_\alpha \cdot \mathbf{k}_\alpha}{k_\alpha} = \frac{\hbar}{2i\mu_\alpha}[2i\mathbf{k}_\alpha] \cdot \frac{\mathbf{k}_\alpha}{k_\alpha} = \frac{\hbar k_\alpha}{\mu_\alpha}. \tag{3.18}$$

The cross section for this scattered wave is therefore just the ratio of the scattered current to the initial current through a unit of solid angle, $\mathrm{d}\Omega$, which is

$$\left(\frac{\mathrm{d}\sigma}{\mathrm{d}\Omega}\right)_{\beta\alpha} = \frac{k_\beta \mu_\alpha}{k_\alpha \mu_\beta}\left|f_{\beta\alpha}(\theta)\right|^2. \tag{3.19}$$

This cross section is often written in terms of the nuclear matrix element, which is defined

$$\mathcal{T}_{\beta\alpha} = \left\langle \varphi_\beta \left| \hat{\mathcal{T}} \right| \varphi_\alpha \right\rangle = \left\langle \varphi_\beta \left| V_\beta \right| \Psi_\alpha \right\rangle, \tag{3.20}$$

where φ_α and φ_β are solutions to the homogeneous version of Eq. 3.12, where the spatial wave function is simply a plane wave [4]. This means that the cross section can be redefined in terms of this matrix element as

$$\left(\frac{\mathrm{d}\sigma}{\mathrm{d}\Omega}\right)_{\beta\alpha} = \frac{\mu_\alpha \mu_\beta}{\left(2\pi\hbar^2\right)^2}\frac{k_\beta}{k_\alpha}\left|\mathcal{T}_{\beta\alpha}\right|^2. \tag{3.21}$$

The dependence of the cross section on the matrix element means that the reaction is more probable for little rearrangement of nucleons between the initial and final wave functions, i.e. when α and β have a large overlap.

At this point, there is still no solution to the scattering problem. The proposed solution in Eq. 3.15 is circular in its definition from Eq. 3.11. Therefore, in order to apply it to transfer reactions, a model must be used.

3.4 The Distorted-Wave Born Approximation

The circular definition in the previous section can be addressed by treating scattering as a perturbation to elastic scattering. Treating elastic scattering as the dominant event when two nuclei collide is called the Born approximation. For elastic scattering, $\beta = \alpha$, allowing the following iterative equation to be defined using Eq. 3.15:

$$\psi_\alpha^{(n+1)}(\mathbf{r}_\alpha) = e^{i\mathbf{k}_\alpha\cdot\mathbf{r}_\alpha} - \frac{e^{ik_\alpha r_\alpha}}{r_\alpha}\frac{\mu_\alpha}{2\pi\hbar^2}\int \mathrm{d}^3\mathbf{r}'\, e^{-i\mathbf{k}_\alpha\cdot\mathbf{r}'}\int \mathrm{d}\xi\ \Phi_\alpha^* V_\alpha \Phi_\alpha \psi_\alpha^{(n)}(\mathbf{r}'), \quad (3.22)$$

where the first term in this expansion is defined as

$$\psi_\alpha^{(0)}(\mathbf{r}_\alpha) = e^{i\mathbf{k}_\alpha\cdot\mathbf{r}_\alpha}. \quad (3.23)$$

This can then be inserted into the expression in Eq. 3.21 to obtain successive terms for calculating the cross section. This series is known as the Born expansion of the scattering wave [4].

However, this doesn't take the effect of the potential of the nucleus on the incoming particle into account. A common way of rectifying this is to use distorted waves instead of plane waves, and this approach is called the distorted-wave Born approximation (DWBA). A similar approach as before is used to obtain the following results; greater detail can be found in Refs. [1, 2, 4].

First, the potential term can be split into two components,

$$V_\beta = U_\beta + \tilde{V}_\beta, \quad (3.24)$$

where U_β is a potential that only depends on the relative coordinates between particles in β. It is also tuned to reproduce the elastic scattering cross section. The term $\tilde{V}_\beta$ is much weaker than U_β, and represents the inelastic channels.

Equation 3.9 can be modified by subtracting U_β from both sides, and then followed by integrating over internal coordinates again

$$\left(E_\beta - \hat{T}_\beta - U_\beta\right)\psi_\beta = \int \mathrm{d}\xi\ \Phi_\beta^*\left(V_\beta - U_\beta\right)\Psi_\alpha. \quad (3.25)$$

From this, a more complex Green's function can be obtained, and this leads to

$$\psi_\beta \to \chi_\alpha^{(\uparrow)}(\mathbf{r}_\alpha)\delta_{\beta\alpha} - \frac{e^{ik_\beta r_\beta}}{r_\beta}\frac{\mu_\beta}{2\pi\hbar^2}\int \mathrm{d}^3\mathbf{r}'\,\chi_\beta^{(\downarrow)*}(\mathbf{r}')\int \mathrm{d}\xi\ \Phi_\beta^*\left(V_\beta - U_\beta\right)\Psi_\alpha^{(\uparrow)}(\mathbf{r}') \quad (3.26)$$

when $\mathbf{r}_\beta \to \infty$. The subscripts $\uparrow$ and $\downarrow$ refer to solutions using incoming and outgoing waves. The functions χ are solutions to the equation

$$\left(E_\gamma - \hat{T}_\gamma - U_\gamma\right)\chi_\gamma = 0, \quad (3.27)$$

for $\gamma = \alpha, \beta$. This can be solved analytically as the potential U_γ is known and it doesn't depend on any internal coordinates. This can be expressed as a summation of spherical harmonics, which has the property

$$\chi^{(\uparrow)}(\mathbf{k}, \mathbf{r}) = \chi^{(\downarrow)*}(-\mathbf{k}, \mathbf{r}). \quad (3.28)$$

The form of Eq. 3.26 closely resembles that of Eq. 3.15, with both containing a spherical wave solution. However, the plane wave in Eq. 3.15 has been replaced by the distorted wave, χ, from Eq. 3.26. This more closely resembles the true scattering process that takes place. The total nuclear matrix element therefore has an additional contribution from the elastic channel element, and becomes

$$\mathscr{T}_{\beta\alpha} = \left\langle e^{ik_\alpha \mathbf{e}_r \cdot \mathbf{r}_\alpha} \left| U_\alpha \right| \chi_\alpha^{(\uparrow)} \right\rangle \delta_{\alpha\beta} + \left\langle \chi_\beta^{(\downarrow)} \Phi_\beta \left| \tilde{V}_\beta \right| \Psi_\alpha^{(\uparrow)} \right\rangle. \tag{3.29}$$

Analogous to the Born expansion, a similar expansion can take place here. However, sufficient accuracy is often reached by truncating the series at the first term, and this is the DWBA. This is obtained by replacing the wave function with one of the distorted waves in the non-elastic channel. The DWBA matrix element is therefore

$$\mathscr{T}_{\beta\alpha}^{\text{DWBA}} = \left\langle \chi_\beta^{(\downarrow)} \Phi_\beta \left| \tilde{V}_\beta \right| \Phi_\alpha \chi_\alpha^{(\uparrow)} \right\rangle. \tag{3.30}$$

This matrix element can be numerically calculated by computer codes. In this work, the codes PTOLEMY [5] and DWUCK5 [6] have been used. The inputs required for these programs will now be discussed.

3.5 Optical Model Potentials

The optical potentials, U_γ, used to model reactions in the DWBA framework are formed from a number of different phenomenological potentials with free parameters that can be used to fit elastic-scattering data. These are deliberately chosen to depend on the magnitude of the separation of the centres-of-mass for the two nucleons in channel γ, so that Eq. 3.27 is satisfied by the optical potential. This is easier to solve, as the many-body problem is then reduced to the interaction between two nuclei. The form of the potentials is often chosen to be a Woods-Saxon potential.

However, the optical model must account for additional open channels as well as the elastic scattering channels. These other channels remove flux from the elastic scattering channel, which is modelled using complex potentials. Combining Eqs. 3.16 and 3.27, and dropping the labels denoting channel, leads to

$$\hbar \nabla \cdot \mathbf{j} = i \psi^* \psi (U^* - U) = 2\rho \, \text{Im}[U], \tag{3.31}$$

which shows that imaginary potentials remove flux from the quantum current [2]. For the reactions in this work, the imaginary potentials define the flux going into inelastic channels.

The total optical potential for a particular partition is

$$U(r) = \left[-U_v f_v(r) + U_C(r) + U_{so} g_{so}(r) \right] + i \left[-W_v f_v(r) + 4W_s \frac{\partial f_s(r)}{\partial r} + W_{so} g_{so}(r) \right], \tag{3.32}$$

where the subscript v denotes volume, C denotes Coulomb, so denotes spin-orbit, and s denotes surface. The strength of the real potentials is labelled U, and the imaginary potentials, W [7, 8]. The function $f_k(r)$ for $k = v, s, so$ is a Woods-Saxon potential of the form,

$$f_k(r) = \frac{1}{1 + \exp\left(\frac{r - R_k}{a_k}\right)}, \tag{3.33}$$

where R_k is the radius of the well, which is $R_k = r_{0,k} A^{\frac{1}{3}}$ in terms of the mass number, A, and the radius parameter for a given channel, $r_{0,k}$; and a_k is the surface diffuseness.

The function $f_k(r)$ also appears in the spin-orbit potential function, $g_{so}(r)$, as a derivative,

$$g_{so}(r) = \left(\frac{\hbar}{m_\pi c}\right)^2 \frac{1}{r} \frac{\mathrm{d} f_{so}(r)}{\mathrm{d}r} \mathbf{L} \cdot \mathbf{S}, \tag{3.34}$$

where m_π is the mass of the pion which mediates the strong nuclear force. This contribution is significant when the two nuclei are very close, and at the nuclear surface. It increases when the total orbital and spin angular momentum components are aligned. In this work, the nuclear spin of the target nucleus is zero, so the total angular momentum used here is entirely dictated by the incoming particle.

Another surface potential appears in the imaginary surface terms, with the factor of 4 used to ensure that the surface term is normalised at $r = R_s$ [7].

Finally, the real Coulomb potential defines the interaction between the electric charge of the two nuclei, and is defined as,

$$U_C(r) = \begin{cases} \frac{Z_1 Z_2 e^2}{4\pi\varepsilon_0} \frac{1}{2R_C} \left(3 - \frac{r^2}{R_C^2}\right), & r \leqslant R_C \\ \frac{Z_1 Z_2 e^2}{4\pi\varepsilon_0} \frac{1}{r}, & r > R_C \end{cases}, \tag{3.35}$$

which is the potential for a uniformly charged sphere of radius R_C and charge $Z_2 e$ acting on a point charge $Z_1 e$. The heavier nucleus is modelled as such, with the radius R_C chosen as the boundary between the nuclear and Coulomb forces dominating [9].

A plot showing the contribution of these different potentials is shown in Fig. 3.4. The sum of these potentials broadly agrees with nuclear behaviour, such as a strong attractive potential near the nuclear centre, with a repulsive Coulomb barrier at larger distances.

While these parameters have been fit to elastic scattering data, the fitting methodology differs between different sets of potentials. Local fits will tailor the optical potentials to a small number of nuclei, which is likely to produce a low-χ^2 fit with little deviation from the data in that region. For global fits, this χ^2 value is likely to

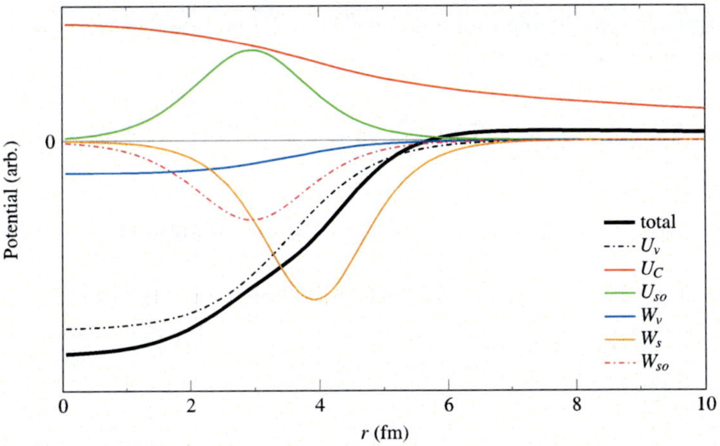

Fig. 3.4 The different components of the optical potential in the exit channel for the ^{28}Mg(d,p)^{29}Mg reaction, with a beam energy of 9.473 MeV/u. U_v has been scaled down by a factor of 8, W_{so} has been scaled up by a factor of 50, and the total has been scaled down by a factor of 4. The fitting parameters have come from the optical model in Ref. [10]

increase when applied to the broader range of nuclei, due to the larger number of data points required to be fitted by this single optical model. This methodology may fit this range reasonably well on average, but be less suitable for some nuclei.

3.6 Spectroscopic Factors

For convenience in this section, the A(d,p)B reaction will be used, though this framework applies to any single-nucleon transfer reaction. The wave functions for the entrance and exit channels for this reaction are therefore

$$\Phi_\alpha = \Phi_\mathrm{A}\Phi_d, \qquad \Phi_\beta = \Phi_\mathrm{B}\Phi_p. \tag{3.36}$$

For the exit channel, a neutron has been removed from the deuteron and added to the nucleus A. As this is a direct reaction, it occurs in a single step. The residual nucleus can therefore be well-described as the target nucleus with a neutron in a single particle state. Therefore,

$$\Phi_\mathrm{B} = \sum_{nj\ell} \gamma_{nj\ell}\zeta_{nj\ell}\Phi_\mathrm{A}, \tag{3.37}$$

where $\zeta_{nj\ell}$ is a neutron single-particle state with principal quantum number n, total angular momentum j, and orbital angular momentum ℓ. The expansion coefficient of the residual nucleus in the single-particle basis set is $\gamma_{nj\ell}$. It describes how much mixing there is of the single-particle states within the nucleus.

In a similar vein, the incoming deuteron wave function can be expressed in terms of the proton's wave function,

$$\Phi_d = \sum_{n'j'\ell'} \gamma_{n'j'\ell'} \zeta_{n'j'\ell'} \Phi_p. \tag{3.38}$$

Using these definitions in Eq. 3.21 with the definition of the matrix element in Eq. 3.30 gives

$$\begin{aligned}\left(\frac{d\sigma}{d\Omega}\right)_{\beta\alpha} &= \frac{\mu_\alpha \mu_\beta}{(2\pi\hbar^2)^2} \frac{k_\beta}{k_\alpha} \left|\left\langle \chi_\beta^{(\downarrow)} \Phi_\beta \left| \tilde{V}_\beta \right| \Phi_\alpha \chi_\alpha^{(\uparrow)} \right\rangle\right|^2 \\ &= \sum_{nj\ell} \sum_{n'j'\ell'} \gamma_{nj\ell}^2 \gamma_{n'j'\ell'}^2 \frac{\mu_\alpha \mu_\beta}{(2\pi\hbar^2)^2} \frac{k_\beta}{k_\alpha} \left|\left\langle \chi_\beta^{(\downarrow)} \zeta_{nj\ell} \Phi_A \Phi_p \left| \tilde{V}_\beta \right| \zeta_{n'j'\ell'} \Phi_A \Phi_p \chi_\alpha^{(\uparrow)} \right\rangle\right|^2.\end{aligned} \tag{3.39}$$

To calculate this, the form of $\tilde{V}_\beta$ has to be specified. In the exit channel, this can be approximated as

$$\tilde{V}_\beta = V_\beta - U_\beta = V_{pA} + V_{pn} - U_\beta \approx V_{pn}, \tag{3.40}$$

where V_{pn} is the interaction potential between the outgoing proton and the transferred neutron, and V_{pA} is the interaction potential between the proton and the target nucleus. By matching U_β to V_{pA}, they can both be ignored, leaving V_{pn} only. This can be calculated numerically, as the two particles involved are independent of the target nucleus. The internal wave functions on either side of the integration are identical, as there is no rearrangement of these nucleons in the transfer, so can be integrated out to give the cross section,

$$\left(\frac{d\sigma}{d\Omega}\right)_{\beta\alpha} = \sum_{nj\ell} \sum_{n'j'\ell'} \gamma_{nj\ell}^2 \gamma_{n'j'\ell'}^2 \underbrace{\frac{\mu_\alpha \mu_\beta}{(2\pi\hbar^2)^2} \frac{k_\beta}{k_\alpha} \left|\left\langle \chi_\beta^{(\downarrow)} \zeta_{nj\ell} \left| V_{pn} \right| \zeta_{n'j'\ell'} \chi_\alpha^{(\uparrow)} \right\rangle\right|^2}_{\text{DWBA cross section}}. \tag{3.41}$$

All terms in the DWBA cross section are now able to be calculated. The unknown terms $\gamma_{nj\ell}^2$ and $\gamma_{n'j'\ell'}^2$ are the spectroscopic factors for the target and ejectile respectively. They are a measure of the overlap between the final state and the initial state with a neutron. Labelling the internal coordinates with ξ again,

$$\gamma_{nj\ell} = \int d\xi\, \zeta_{nj\ell}^* \Phi_A^* \Phi_B, \tag{3.42}$$

and similarly for $\gamma_{n'j'\ell'}$.

For this work, the spectroscopic factor can be further simplified, as all the reactions of interest are on even-even nuclei with spin zero. The angular momentum is therefore the same for both the ejectile and the residual nucleus. Equation 3.41 can then be written

$$\left(\frac{\mathrm{d}\sigma}{\mathrm{d}\Omega}\right)^{(\mathrm{EXP})}_{\beta\alpha} = S^B_{j\ell} S^b_{j'\ell} \left(\frac{\mathrm{d}\sigma}{\mathrm{d}\Omega}\right)^{(\mathrm{DWBA})}_{\beta\alpha}, \tag{3.43}$$

where $S^B_{j\ell} = \gamma^2_{nj\ell}$ is the spectroscopic factor for the residual nucleus, and $S^b_{j'\ell} = \gamma^2_{n'j'\ell}$ is the spectroscopic factor for the ejectile [8, 11]. This assumes the reaction occurs in a single step, as this allows the inert core and nucleon model to be used.

3.7 Bound-State Parameters

The bound-state wave functions of both target and projectile are used to calculate the DWBA cross section. In the previous section, this was contained in the single-particle wave functions ζ and spatial wave functions χ for the entrance and exit channels.

These are calculated by assuming that these wave functions are eigenfunctions of a Woods-Saxon well with a form described by the real part of Eq. 3.32. The depth of the potential well is then varied so that the energy eigenvalue corresponds to the binding energy of the populated state by the transferred nucleon [12]. Following from the work of Refs. [13, 14], the potentials used for the target had a Woods-Saxon radius parameter $r_{0,v} = 1.28$ fm, and a diffuseness parameter $a_v = 0.65$ fm. The spin-orbit potential had $r_{0,so} = 1.1$ fm, $a_{so} = 0.65$ fm, and a depth of $U_{so} = 6$ MeV.

3.8 Unbound States

The discussion of states thus far has focused on bound states. However, unbound states may also be populated—especially for light nuclei. This presents some difficulties in calculating the DWBA cross section. The wave functions for unbound states are markedly different, as illustrated in Fig. 3.5. This shows the wave functions for states in a finite square well, which is an approximation to the Woods-Saxon potential for the nucleus. For bound states, the wave function is oscillatory within the nucleus and exponentially decays in the region beyond the nuclear surface. This exponential contribution increases as the binding decreases. Once the nucleon is unbound, the wave function outside the nuclear surface is also oscillatory. While this oscillation amplitude will decrease as $r \to \infty$, this happens more slowly than any exponential tails of the bound-state wave functions.

Therefore, computing radial integrals numerically for unbound states becomes difficult, and naïvely requires integrating the wave function at distances far outside the nuclear surface in order to account for these oscillations. This is computationally taxing. Huby and Mines [16] proposed introducing a convergence factor of $e^{-\alpha r}$ into the wave function in order to make this easier, and then extrapolate to $\alpha \to 0$. This still requires numerical integration at relatively large distances, however, so is not ideal.

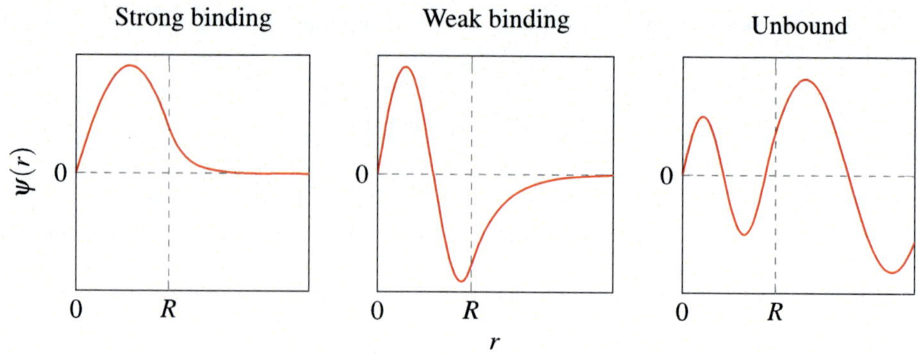

Fig. 3.5 Wave functions for a finite square well of radius R. This illustrates the difference in wave function for bound and unbound states in a nucleus [15]

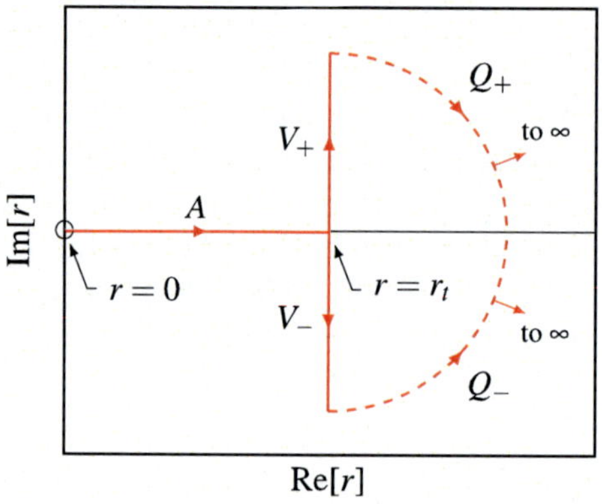

Fig. 3.6 An outline of the steps taken to change the pure real integral used in calculating spectroscopic factors for bound states into a complex contour integral to deal with unbound states. Adapted from Fig. 3 in Ref. [17]

The solution used in this work was first proposed by Vincent and Fortune [17] and is implemented in the code DWUCK5, but not in PTOLEMY. Their method is illustrated in Fig. 3.6. Instead of integrating to large distances along the real axis, the integral is split at $r = r_t$, a point where the nuclear potentials can be considered negligible. For $r \leqslant r_t$, the radial integral is carried out as before along the real axis. For $r > r_t$, the wave function is split into outgoing and incoming components of the wave. The outgoing parts are integrated in the region $\mathrm{Im}[r] \geqslant 0$, and the incoming parts are integrated in the region $\mathrm{Im}[r] \leqslant 0$.

For each of the waves in the wave function, they are integrated along the line $V_\pm$, and then along an arc $Q_\pm$ joining the end of this line to the real axis with a constant radius centred at $r = r_t$. The radius of each arc is then taken in the limit where they tend to infinity, and the contribution from the integration along $Q_\pm \to 0$.

Additionally, the integrands along $V_\pm$ tend to 0 exponentially at large $\mathrm{Im}[r]$, though does exhibit some oscillations. Therefore, instead of truncating a real integral at large r, leading to large uncertainties in the integration, it is better to do this contour integration so that the truncation error is minimised.

In addition to treating the DWBA calculations differently, the unbound states, also have some width to them. This width Γ is proportional to the spectroscopic factor, and is

$$\Gamma = C^2 S \Gamma_{sp}, \tag{3.44}$$

where Γ_{sp} is the pure single-particle width calculated from the optical potential [18, 19].

3.9 Sum Rules

Residual interactions, as described in Sect. 2.2, spread the independent-particle model (IPM) orbital strength across several states. The spectroscopic factors quantify what proportion of a given IPM state in the residual nucleus is in a particular single-particle orbital. If all of these states in the nucleus have been measured, then the total occupancies and vacancies of particular orbitals can be extracted using the Macfarlane-French sum rules [20].

Dropping the entrance and exit channel labels, Eq. 3.43 can be rewritten as

$$\left(\frac{\mathrm{d}\sigma}{\mathrm{d}\Omega}\right)_{\mathrm{EXP}} = C^2 S_{j\ell} \left(\frac{\mathrm{d}\sigma}{\mathrm{d}\Omega}\right)_{\mathrm{DWBA}}, \tag{3.45}$$

where $S_{j\ell}$ is the spectroscopic factor for the particular state, and C^2 is the isospin-coupling Clebsch-Gordan coefficient. The term $C^2 S_{j\ell}$ can be extracted simply by dividing the experimental and DWBA cross sections. The Macfarlane-French sum rules then state that the number of vacancies, N_V, is

$$N_V = \sum (2j+1) C^2 S_+, \tag{3.46}$$

where the sum is over the possible transitions where j was the total angular momentum transferred [21]. The spectroscopic factors $C^2 S_+$ can be obtained using nucleon-addition reactions. This has been defined so that for an empty orbital,

$$\sum C^2 S_+ = 1, \tag{3.47}$$

so that $N_V = 2j + 1$, the total number of vacancies in the orbital.

Similarly, the number of occupancies is

$$N_O = \sum C^2 S_-, \tag{3.48}$$

where $C^2 S_-$ is obtained using nucleon-removal reactions. For a full orbital,

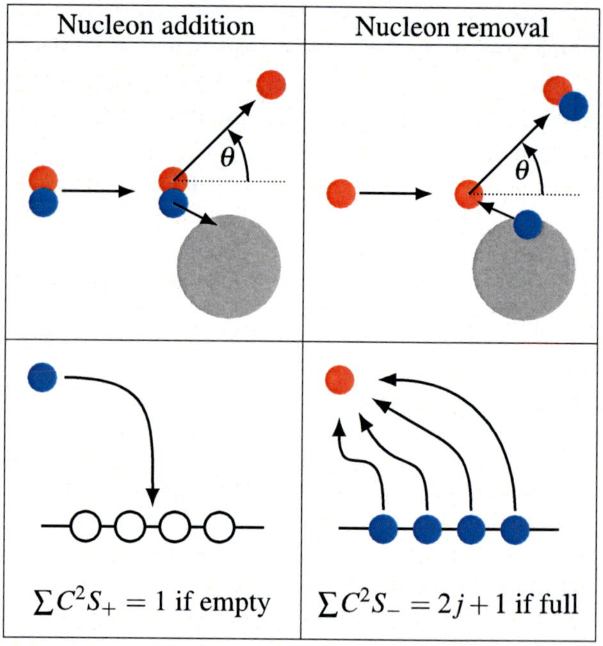

Fig. 3.7 An illustration of the Macfarlane-French sum rules using (*d*,*p*) and (*p*,*d*) reactions

$$\sum C^2 S_- = 2j + 1, \tag{3.49}$$

the total number of nucleons in the orbital. These rules are represented pictorially in Fig. 3.7 with (*d*,*p*) and (*p*,*d*) reactions.

Summing the number of occupancies and vacancies should give the degeneracy of the orbital of $2j + 1$. Experimentally, this has shown to be affected by a quenching factor that approximately halves this value [14]. This is due to short-range interactions that move strength to high-excitation energies which are not typically probed in transfer reaction experiments.

References

1. Satchler GR (1990) Introduction to nuclear reactions. Palgrave Macmillan, London
2. Glendenning NK (1983) Direct nuclear reactions. Elsevier
3. Arfken G (1985) Mathematical methods for physicists, 3rd edn. Academic
4. Messiah A (1962) Quantum mechanics, vol II (trans: Potter J). North-Holland Publishing Company, Amsterdam [in French]
5. Macfarlane MH, Pieper S. Argonne National Laboratory (unpublished)
6. Kunz PD. DWUCK5. Computer code (unpublished)
7. Hodgson PE, Gadioli E, Gadioli-Erba E (1997) Introductory nuclear physics. Oxford University Press
8. Szwec SV (2017) Transfer reaction studies of medium mass nuclei—single-particle occupancies and neutrinoless double beta decay. PhD Thesis, The University of Manchester. EThOS: 740351

9. Bertulani CA (2007) Nuclear physics in a nutshell, vol 2. Princeton University Press
10. Koning AJ, Delaroche JP (2003) Local and global nucleon optical models from 1 keV to 200 MeV. Nucl Phys A 713(3):231–310
11. Sharp DK (2012) Trends in single-particle energies in N = 51 nuclei. PhD Thesis, The University of Manchester. EThOS: 559344
12. Wildenthal BH, Newman E, Auble RL (1971) Study of the level structure of N = 82 nuclei via proton-transfer reactions. Phys Rev C 3:1199–1220
13. Kramer GJ, Blok HP, Lapikás L (2001) A consistent analysis of (e;e0p) and (d;3He) experiments. Nucl Phys A 679(3):267–286
14. Kay BP, Schiffer JP, Freeman SJ (2013) Quenching of cross sections in nucleon transfer reactions. Phys Rev Lett 111:042502
15. Phillips AC (2003) Introduction to quantum mechanics. Wiley
16. Huby R, Mines JR (1965) Distorted-wave born approximation for stripping to virtual levels. Rev Mod Phys 37:406–408
17. Vincent CM, Fortune HT (1970) New method for distorted-wave analysis of stripping to unbound states. Phys Rev C 2:782–792
18. Brown SM (2010) Neutron shell breaking in neutron-rich neon isotopes. PhD thesis, University of Surrey. EThOS: 529440
19. Cooper SG, Huby R, Mines JR (1982) Prescription for calculating stripping to states just unbound or very weakly bound. J Phys G: Nucl Phys 8(4):559–566
20. Macfarlane MH, French JB (1960) Stripping reactions and the structure of light and intermediate nuclei. Rev Mod Phys 32:567–691
21. Schiffer JP, Hoffman CR, Kay BP, Clark JA, Deibel CM, Freeman SJ, Howard AM, Mitchell AJ, Parker PD, Sharp DK et al (2012) Test of sum rules in nucleon transfer reactions. Phys Rev Lett 108:022501

Chapter 4
Experimental Details for the $d(^{28}Mg,p)^{29}Mg$ Reaction at ISOLDE, CERN

This chapter outlines the experimental details for measuring the cross sections to excited states in ^{29}Mg in reactions induced by a beam of ^{28}Mg used to bombard a deuterated target in the ISOLDE Solenoidal Spectrometer (ISS). ISS has been constructed specifically to study direct reactions in inverse kinematics. This chapter will discuss the production and acceleration of the radioactive ion beam at the ISOLDE facility, CERN, and the principles behind ISS.

4.1 The ISOLDE Facility

ISOLDE, an online isotope separator, is a nuclear physics facility at CERN that specialises in the production of radioactive ion beams. A diagram of the ISOLDE facility layout is shown in Fig. 4.1.

4.1.1 Enriched Target, HRS, and RILIS

The protons from the proton synchrotron booster (PSB) [2, 3] are delivered to the ISOLDE target station where they impact on a target, creating a large number of different atomic nuclei. This can occur via a number of different reaction pathways, such as spallation, fragmentation, or fission [4].

A silicon carbide (SiC) target [5] was used to create the ^{28}Mg beam owing to a higher production yield. The target is heated to promote the release of the nuclear fragments through diffusion and surface desorption, and has a maximum operating temperature of 1923 K [6].

The atomic nuclei diffuse out of the target, and are subsequently ionised using the Resonance Ionisation Laser Ion Source (RILIS) [7]. Using a predetermined ionisation scheme, two lasers of different frequencies are used to resonantly excite the desired

P. T. MacGregor, *Single-Particle Structure of ^{29}Mg on the Approach to the N=20 Island of Inversion*, Springer Theses, https://doi.org/10.1007/978-3-031-19119-0_4

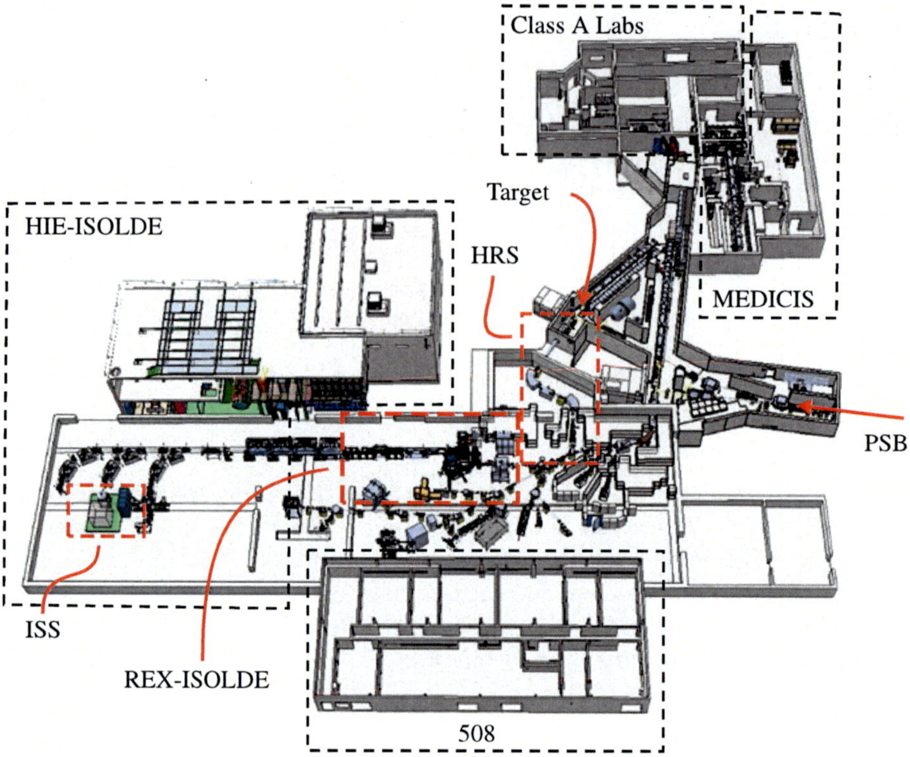

Fig. 4.1 A schematic of the ISOLDE facility. The relevant sections of the beam line have been highlighted, and will be discussed. Adapted from Ref. [1]

element to subsequent excited atomic states. A third laser is used to ionise the atom to a 1+ charge state. As the atomic energy levels of elements have distinct energy gaps, this series of precisely tuned lasers are able to ionise the desired element with few excitations of additional unwanted elements.

However, isobaric beam contamination remains at this stage, so the ions are then accelerated into a mass separator at energies between 30 and 60 keV. While ISOLDE has two mass separators, the high-resolution separator (HRS) [8] was used for this experiment. This has a theoretical mass resolving power of 15,000 [9, 10], and separates a given A/q from the rest of the fragments. After mass-separation, the ions are directed into the next stage of acceleration in REX-ISOLDE.

4.1.2 REX-ISOLDE

REX-ISOLDE (Radioactive beam EXperiment at ISOLDE) [11, 12] is used to accelerate the nuclear fragments further. A diagram of the components are shown in Fig. 4.2.

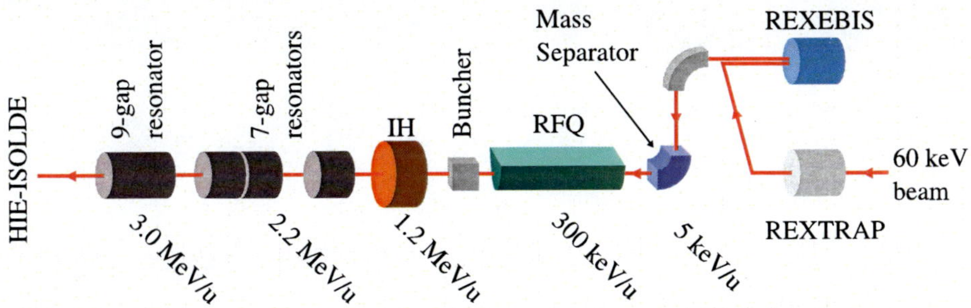

Fig. 4.2 The REX-ISOLDE portion of the beam line. Adapted from Refs. [13, 14]

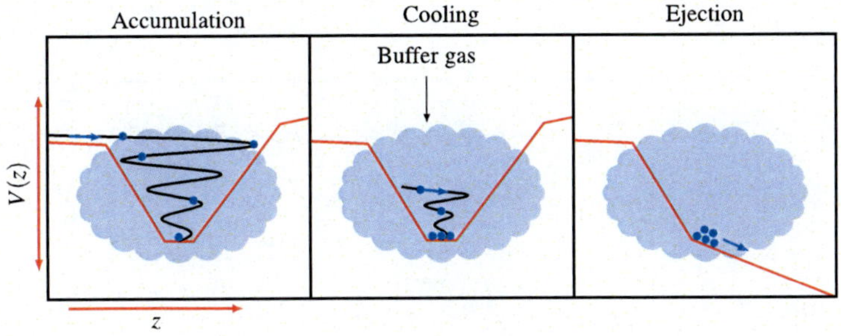

Fig. 4.3 An illustration of the principles behind REXTRAP. Adapted from Ref. [15]. The vertical axis denotes the energy of each particle within the trap, and the horizontal axis is the beam axis

First, the beam enters REXTRAP, which is a Penning trap that accumulates, cools, and bunches the beam. An illustration of the function of REXTRAP is shown in Fig. 4.3.

It consists of a 1 m cylindrical trap that has a 3 T solenoidal magnetic field and is filled with a buffer gas, such as neon or argon. Ions are slowed by a 60 kV potential at the entrance, so that they enter the trap with electronvolt energies. The gas further slows them by friction through collisions. Upon reaching the other side of the cylinder, they reach a potential wall and are held in the trap to bunch further and cool. They are then deflected out of the trap by changing the electric potential at the exit [16, 17]. The cooling and bunching helps improve the signal-to-background ratio for low-intensity beams, and is necessary for entering the next stage.

Upon exiting REXTRAP, the ions are directed towards the REXEBIS (REX Electron Beam Ion Source). This is a 0.8 m long trap inside a 1.2 m long solenoid. The efficiency of ions entering the REXEBIS is strongly linked to the effectiveness of the cooling in REXTRAP; the cooling time must be at least 10 ms [18]. An illustration of the function of an EBIS is shown in Fig. 4.4.

Once inside REXEBIS, the ions are held by an electrostatic barrier, confining the ions and preventing escape. An electron beam is fired at the ions. This beam has a high current density (of the order 100 A cm^{-2}), such that there is a high probability of collisions with the ions. This causes the ions to be further ionised stepwise forming

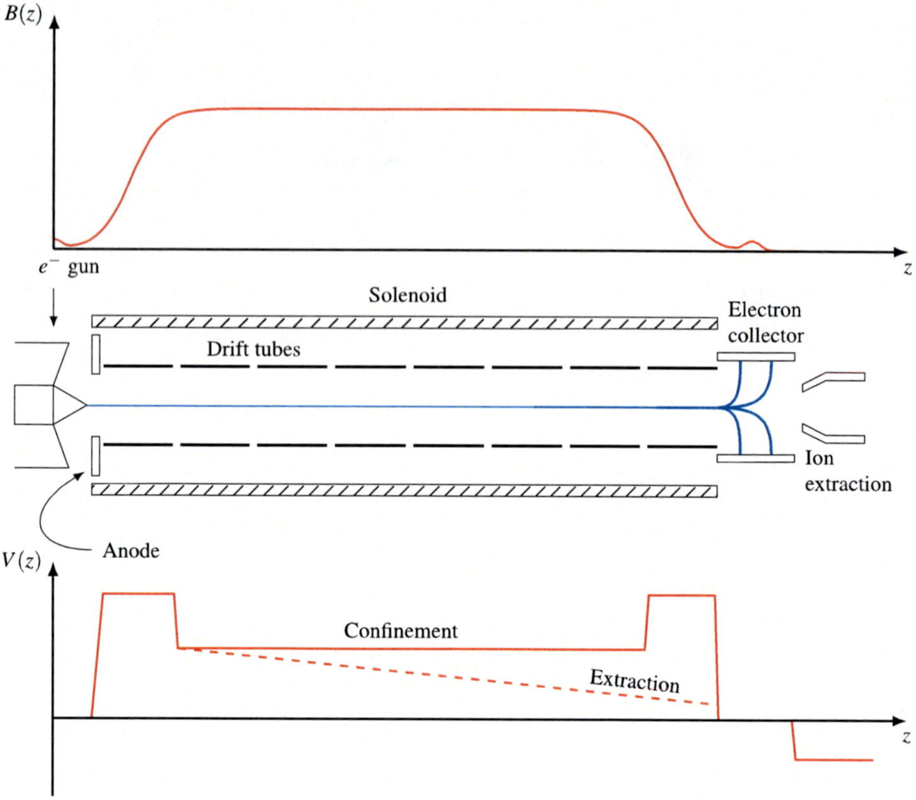

Fig. 4.4 An illustration of the function of an EBIS. The axial magnetic field is shown at the top, a schematic of the interior is shown in the middle, and the axial electric potential is shown for both extraction and confinement of ions. Adapted from Ref. [13]. The ion extraction is placed at the opposite end of the EBIS here for clarity; in reality, it is at the entrance for REXEBIS

a distribution of charge states. The electrostatic barrier is then lowered, and the ions leave through the same side, where they are directed towards a separator. This separates the ions based on their mass-to-charge ratio, which helps to remove any contamination from the buffer gas and to select the desired charge state [13, 18, 19]. For this experiment, the REXEBIS strips electrons from $^{28}Mg^{+}$ to become $^{28}Mg^{9+}$.

The ions are then directed into the REX-LINAC, which was shown in Fig. 4.2. The first component is a radiofrequency quadrupole (RFQ), which accelerates the beam to 0.3 MeV/u in the space of 3 m using quadrupole electric-fields with a resonance frequency of 101.28 MHz. The second component is a buncher, which re-bunches the beam in 1.7 m to help it enter the third component, which is the interdigital H-type (IH) structure. This is a drift-tube LINAC, and accelerates the beam to 1.2 MeV/u in 1.5 m at the same frequency [20].

The beam then passes through three 7-gap spiral resonators, which can accelerate the beam to 2.2 MeV/u. This also uses the gaps between drift tubes to accelerate the beam, but the drift tubes are mounted on spirals, where changing the spiral length and shape allows the properties of the accelerator to be manipulated—see Refs. [21, 22] for details. The beam finally enters the 9-gap IH resonator, which accelerates the beam to a maximum of 3 MeV/u. The beam then passes into the HIE-ISOLDE accelerator modules.

4.1.3 HIE-ISOLDE LINAC

HIE-ISOLDE (High Intensity and Energy ISOLDE) refers to energy upgrades carried out at the ISOLDE facility. This primarily consists of four superconducting cryomodules. The first was installed in 2015, with subsequent cryomodules installed every year until 2018 [23]. A diagram of HIE-ISOLDE at the time of the $d(^{28}\text{Mg},p)^{29}\text{Mg}$ experiment is shown in Fig. 4.5.

Each cryomodule operates at a temperature of 4.5 K, and has a length of 2.3 m [25]. Each high-energy cryomodule contains five superconducting cavities for acceleration and a superconducting solenoid for focusing the beam [26]. This is designed to give a minimum beam energy of 9.2 MeV/u for $2.5 \leqslant A/q \leqslant 4.5$, and allows beams with $A/q = 4$ to reach 10 MeV/u, the desired final beam energy at the end of the upgrade process [23].

For this experiment, the beam energy was 9.473 MeV/u with a FWHM energy spread of 0.3% [27]. This was the maximum beam energy that HIE-ISOLDE was able to produce for this A/q at the time of the experiment. The typical PSB proton current was 1.75 μA, which corresponded to 1.2×10^6 pps of $^{28}\text{Mg}^{9+}$ at ISS [5].

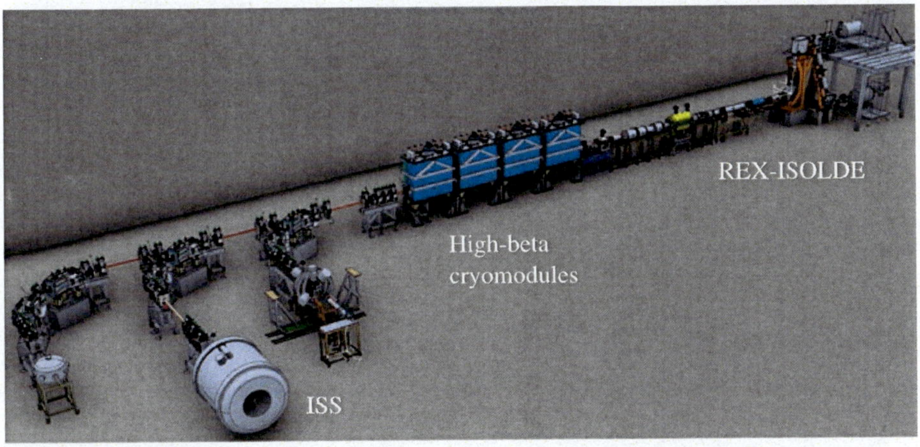

Fig. 4.5 Diagram of HIE-ISOLDE with ISS installed. Adapted from Refs. [9, 24]

4.2 Inverse Kinematics

The description of transfer reactions given in Sect. 3.2 is for cases involving a light-ion beam incident on a target containing heavy nuclei. This is known as normal kinematics (NK). For reactions involving short-lived species such as ^{28}Mg (half-life $\sim$ 21 h [28]), a radioactive beam is used in inverse kinematics (IK), where the heavier nuclei of interest are used to bombard a lighter target.

Transfer reactions in NK and IK are indistinguishable in the centre-of-momentum (CM) frame. In the lab frame, however, there are noticeably different characteristics. A diagram of the two different frames is shown in Fig. 4.6 for IK.

Here, particles 1 and 2 are the heavy beam and light target particles respectively. Particles 3 and 4 are the light and heavy ejectiles respectively. The terms v and u denote velocities in the CM and lab frame respectively, and θ_{cm} and θ_{lab} define the directions of the beams of ejectiles in each frame.

The angle of the scattered particle is conventionally defined relative to the beam direction, as θ_{cm} is here. However, it is often more convenient to use $\eta_{cm} = \pi - \theta_{cm}$, which is directly comparable to θ_{cm} in NK, as η_{cm} in IK and θ_{cm} in NK correspond to a shallow angle of scattering of the ejectile in the CM frame.

The following will consider classical kinematics, which is sufficiently accurate for the experiment in question. However, a relativistic treatment is presented in Appendix A.

The excitation energy of the residual nucleus for a given reaction, beam energy, and scattering angle can be extracted by measuring the kinetic energy of the light ejectile. The kinematics of the particles in the lab and CM frames show how these two quantities are related. The initial kinetic energy in the CM frame is

$$T_i^{cm} = \frac{1}{2}m_1v_1^2 + \frac{1}{2}m_2v_2^2 = \frac{m_2}{m_1+m_2}T_1, \tag{4.1}$$

where $T_1 = \frac{1}{2}m_1u_1^2$, which is the beam energy in the lab frame. This assumes that the target is stationary in the lab frame so that $u_2 = 0$. The final kinetic energy in the CM frame is

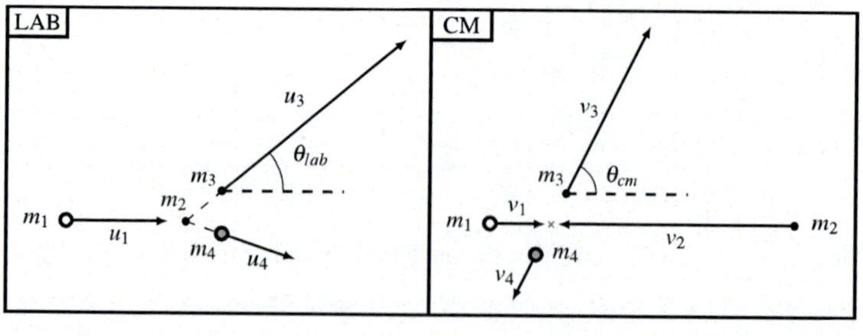

Fig. 4.6 The lab frame and the CM frame for a reaction in IK. Definitions of symbols are in the body text

$$T_f^{cm} = \frac{1}{2}m_3v_3^2 + \frac{1}{2}m_4v_4^2 = T_i^{cm} + Q - E_x, \tag{4.2}$$

where E_x is the excitation energy of the residual nucleus, and Q is the ground-state Q value for the reaction,

$$Q = [(m_1 + m_2) - (m_3 + m_4)]\, c^2. \tag{4.3}$$

The velocity of the light ejectile in the CM frame is therefore

$$v_3 = \sqrt{\frac{2m_4(T_i^{cm} + Q^*)}{m_3(m_3 + m_4)}}, \tag{4.4}$$

where $Q^* = Q - E_x$. This CM velocity can be transformed into a lab velocity using the cosine rule between θ_{lab} and θ_{cm}, and the velocity of the centre of momentum,

$$v_{cm} = \frac{m_1u_1}{m_1 + m_2} = \frac{\sqrt{2T_1m_1}}{m_1 + m_2}. \tag{4.5}$$

This relationship is represented graphically in Fig. 4.7.

Thus,

$$u_3^2 = v_3^2 + v_{cm}^2 - 2v_3v_{cm}\cos(\pi - \theta_{cm}), \tag{4.6}$$

and therefore

$$T_3 = \frac{1}{2}m_3u_3^2 = \frac{1}{2}m_3v_3^2 + \frac{1}{2}m_3v_{cm}^2 + m_3v_3v_{cm}\cos\theta_{cm}. \tag{4.7}$$

The measured energy of the ejectile particle, T_3, can therefore be directly linked to E_x using Eqs. 4.4 and 4.7. Using Eqs. 4.1, 4.4 and 4.5, $\frac{v_3}{v_{cm}}$ is written as

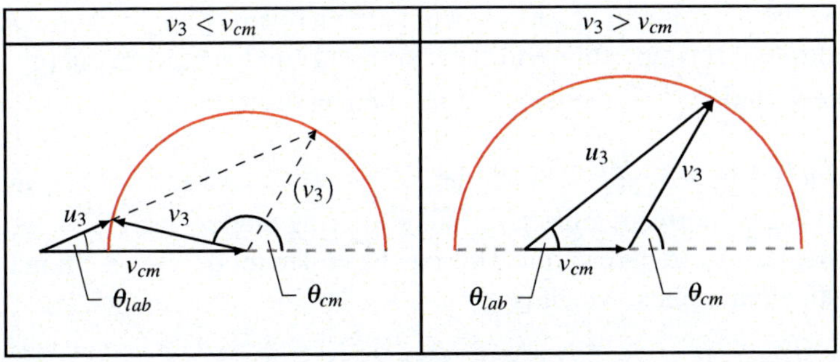

Fig. 4.7 The relationship between u_3 and v_3 in two distinct cases. The red arc drawn in both cases represents the locus of the head of the vector v_3, where $0 \leqslant \theta_{cm} \leqslant \pi$

$$\frac{v_3}{v_{cm}} = \sqrt{M\left(\frac{m_1+m_2}{m_3+m_4}\right)\left(1+\frac{Q^*}{T_i^{cm}}\right)}, \tag{4.8}$$

where $M = \frac{m_2 m_4}{m_1 m_3}$. This fraction is fixed for a given value of Q^*, but describes three different kinematic cases depending on its value. The quantity M plays the most significant role in determining which case the reaction takes.

The first case is when $v_3 < v_{cm}$. This occurs for removal reactions, such as (*p*,*d*), where $M < 1$. It also occurs in addition reactions where E_x is comparable in size to T_i^{cm}. In this case, θ_{cm} always has two values for a given θ_{lab}, except at some maximum value. Ejectiles in the lab frame always have a component of velocity in the same direction as the beam.

The second case is when $v_3 > v_{cm}$. This occurs for addition reactions, such as (*d*,*p*), for smaller E_x where $M > 1$.[1] In this case, there is a single value of θ_{lab} for every θ_{cm}. There is also the possibility of ejectiles having a component of velocity antiparallel to the beam. If v_3 is similar in magnitude to v_{cm}, then θ_{lab} depends strongly on θ_{cm}. If $v_3 \gg v_{cm}$, where $M \gg 1$ (as is the case in normal kinematics), then $\theta_{lab} \approx \theta_{cm}$.

The third case, not illustrated in Fig. 4.7, is where $v_3 = v_{cm}$ because $Q = E_x = 0$. This is elastic scattering and the nuclides remain the same before and after scattering. In this case, θ_{lab} is still single-valued, but is bounded and has a maximum at $\frac{\pi}{2}$.

4.3 Challenges in Inverse Kinematics

In IK, there are kinematic effects that limit the energy resolution of the measured ejectile energy. These can be seen in Fig. 4.8, which shows the change in T_3 for a given θ_{lab} for both NK and IK. These effects make measurements in IK more challenging than in NK in terms of resolving states of interest in the residual nucleus.

The first effect is kinematic shift (KS), which can be understood with reference to the variation of T_3 with θ_{lab} as illustrated in Fig. 4.8 for NK and IK. Any detector measuring the outgoing light particle will have an opening angle, $\Delta\theta_{lab}$. Any particle passing through this aperture will have an associated spread in energy, $\frac{\mathrm{d}T_3}{\mathrm{d}\theta_{lab}}\Delta\theta_{lab}$, due to the variation of T_3 with θ_{lab}. This contributes to the energy resolution of the detector.

As seen in Fig. 4.8, $\frac{\mathrm{d}T_3}{\mathrm{d}\theta_{lab}}$ is much larger (as much as a factor of 5) in IK than in NK at lab angles corresponding to small scattering angles, which correspond to the highest yield for a direct reaction. This can be understood by examining the gradient of these kinematic lines, which is

[1] $v_3 > v_{cm}$ in addition reactions until $E_x = T_i^{cm}\left(1 - \frac{m_1 m_3}{m_2 m_4}\right) + Q$, which is approximately 60% of T_i^{cm} (10.6 MeV) for the $d(^{28}Mg,p)^{29}Mg$ reaction.

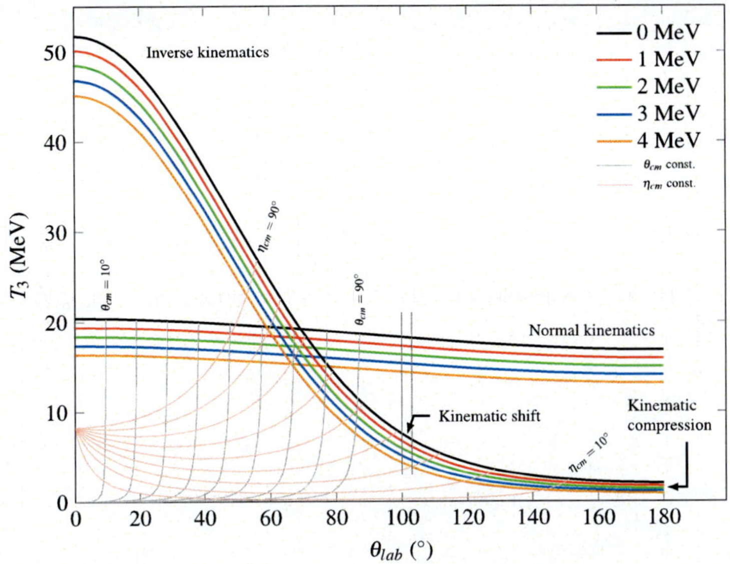

Fig. 4.8 The ejectile energy T_3 for equally spaced excitation energies for the ^{28}Mg(*d*,*p*) reaction in IK and NK. The angle η_{cm} is used to label the analogous region in IK to the region of small θ_{cm} in NK. Constant values of these CM angles have been drawn as lines in 10° increments for NK and IK respectively, with the start and end lines labelled

$$\frac{\mathrm{d}T_3}{\mathrm{d}\theta_{lab}} = -m_3 v_3 v_{cm} \sin\theta_{cm} \frac{\mathrm{d}\theta_{cm}}{\mathrm{d}\theta_{lab}}. \tag{4.9}$$

The derivative $\frac{\mathrm{d}\theta_{cm}}{\mathrm{d}\theta_{lab}}$ can be neglected for both IK and NK, as it does not have a very significant impact upon the gradient.[2] As T_i^{cm} is a constant for a reaction in NK and IK, v_3 and m_3 are identical for a given reaction in NK and IK. Therefore, the controlling factor for the gradient is v_{cm}. Rewriting Eq. 4.5 in terms of T_i^{cm} from Eq. 4.1 yields

$$v_{cm} = \sqrt{\frac{2m_1 T_i^{cm}}{m_2(m_1 + m_2)}}. \tag{4.10}$$

As $m_1 > m_2$ in IK, v_{cm} is larger and so $\frac{\mathrm{d}T_3}{\mathrm{d}\theta_{lab}}$ is larger in IK. Therefore, KS is more significant in IK.

The second important effect in IK is kinematic compression (KC), which can be understood with reference to the spacing in T_3 for different populated states in the residual nucleus for IK and NK in Fig. 4.8. It is crucial to have ejectiles with well-separated energies in order to resolve states in the excitation-energy spectrum.

[2] A complete derivation of this factor can be found in Appendix B.

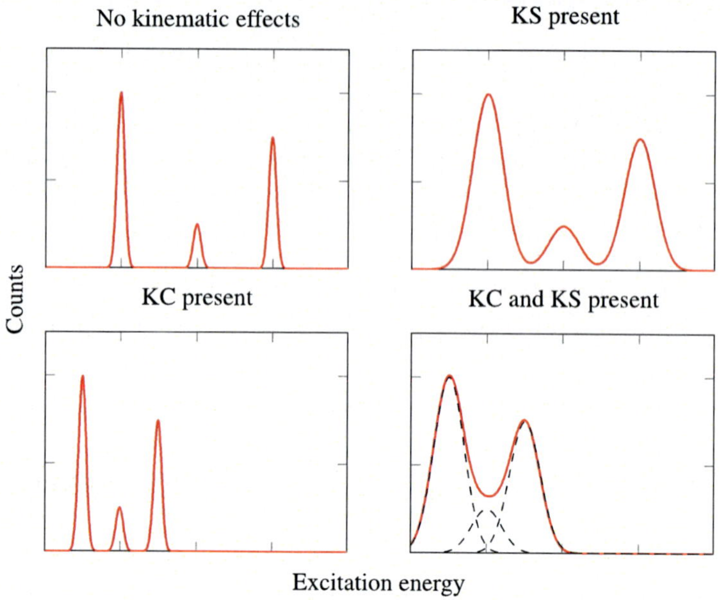

Fig. 4.9 An illustration of an excitation-energy spectrum with and without the effects of KS and KC

In IK, states separated by a given excitation energy are compressed together more than in NK. This can be demonstrated by calculating the difference in ejectile energy for two excited states in NK and IK. This difference, ΔT_3, is

$$\Delta T_3 = A + B\sqrt{\frac{m_1}{m_2}}\cos\theta_{cm} = A - B\sqrt{\frac{m_1}{m_2}}\cos\eta_{cm}, \tag{4.11}$$

where A and B are positive constants in IK and NK.[3]

This relationship shows that both IK and NK will experience KC as θ_{cm} increases, as shown in Fig. 4.8. The degree of this compression is dictated by the factor $\sqrt{\frac{m_1}{m_2}}$, which is large in IK and small in NK. States in NK are less affected by the compression for forward angles (small θ_{cm}), whereas IK are greatly affected at forward angles (small η_{cm}).

Both KS and KC are both present at these forward angles in the CM frame, which is a typical region of high yield in transfer-reaction experiments [29]. The cumulative effect of KS and KC is illustrated in Fig. 4.9.

The top left spectrum shows three distinct peaks. The effect of KS (top right) broadens the peaks, and the effect of KC (bottom left) forces the peaks together. The overall effect (bottom right) can obscure weaker peaks, and makes the task of resolving peaks in the excitation-energy spectrum much harder.

[3] Specifically, for excited states with excitation energies $E_x^{(1)}$ and $E_x^{(0)}$, the constant $A = \frac{m_4}{m_3+m_4}\left(E_x^{(1)} - E_x^{(0)}\right)$ and $B = 2\sqrt{\frac{m_3 m_4 T_i^{cm}}{(m_1+m_2)(m_3+m_4)}}\left[\sqrt{T_i^{cm} + Q - E_x^{(0)}} - \sqrt{T_i^{cm} + Q - E_x^{(1)}}\right]$.

These kinematic effects create issues for conventional experimental set-ups where detectors are arranged at fixed angles. Detectors require excellent angular and energy resolution in order to distinguish ejectiles, so that yields can be extracted. Ejectile energies for the desired angular range are also much smaller in IK than in NK. The resulting poor excitation-energy resolution can be ameliorated to some degree by using high-resolution γ-ray detector arrays, where the detected ejectile and γ-rays emitted by the daughter nucleus occur in coincidence. This is the method used by the MINIBALL experiment at ISOLDE for example [30]. However, γ-ray detection is often limited to efficiencies of a few percent and thicker targets are required, worsening the ejectile energy resolution. Such techniques are also limited to states that decay via γ-rays and are complicated in the presence of longer-lived isomeric decays to obtain reasonable data rates.

The ISOLDE Solenoidal Spectrometer is designed to overcome the kinematic effects that hinder resolution in IK, in particular KC.

4.4 The Solenoid Concept

The ISOLDE Solenoidal Spectrometer design is based upon HELIOS (HELIcal Orbital Spectrometer) at Argonne National Laboratory, USA [31]. They are both made from repurposed MRI magnets. A schematic diagram of ISS is shown in Fig. 4.10 for this particular experiment.

Reactions are performed within the solenoid, with the beam parallel to the magnetic field axis. The ^{28}Mg travels down the beam line towards a deuterated target. In the case of the *(d,p)* reaction, the protons corresponding to the peak of the transfer yield are emitted into the backward hemisphere in helical orbits, and return to the axis to hit an array of position-sensitive silicon detectors.

4.4.1 Kinematics in a Solenoid Field

An ejectile from a reaction of charge q in a uniform magnetic field $\mathbf{B}$ experiences a centripetal force perpendicular to the magnetic field and its velocity $\mathbf{u_3}$, which can be expressed as

$$q\mathbf{u_3} \times \mathbf{B} = qu_\perp B = \frac{m_3 u_\perp^2}{r}, \tag{4.12}$$

where $u_\perp$ is the velocity component of $\mathbf{u_3}$ perpendicular to the magnetic field, and r is the radius of the helical orbit. This radius is therefore

$$r = \frac{m_3 u_\perp}{qB}. \tag{4.13}$$

A diagram of this helical orbit within the solenoid is shown in Fig. 4.11.

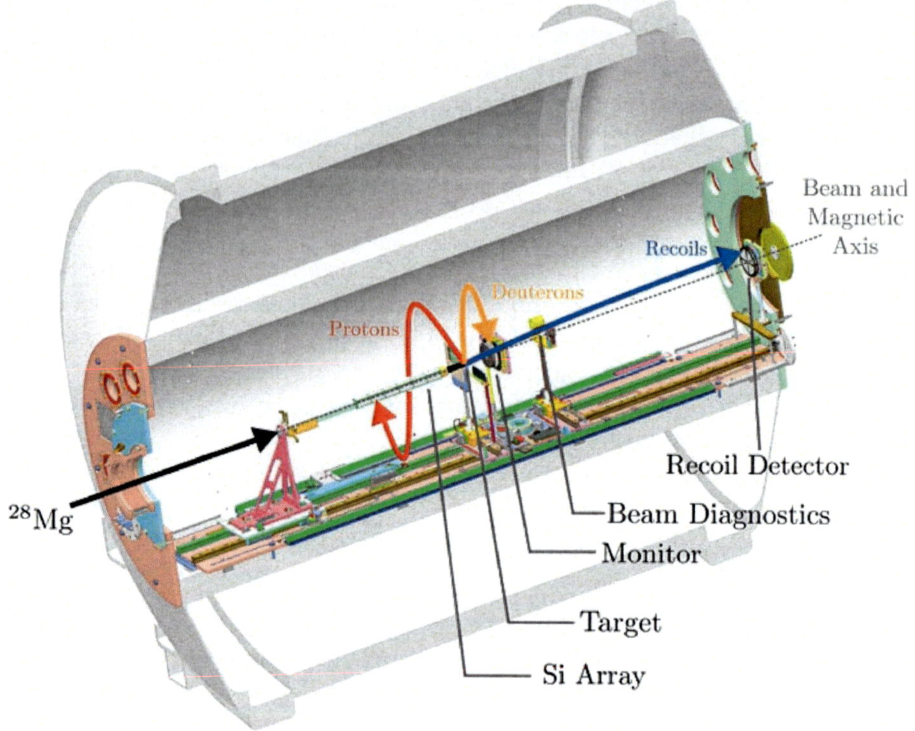

Fig. 4.10 A cross section of the inside of ISS, with detectors arranged for the $d(^{28}Mg,p)^{29}Mg$ reaction. The magnetic field is parallel to the beam-line axis. The trajectories of the different particles are shown hitting their respective detectors

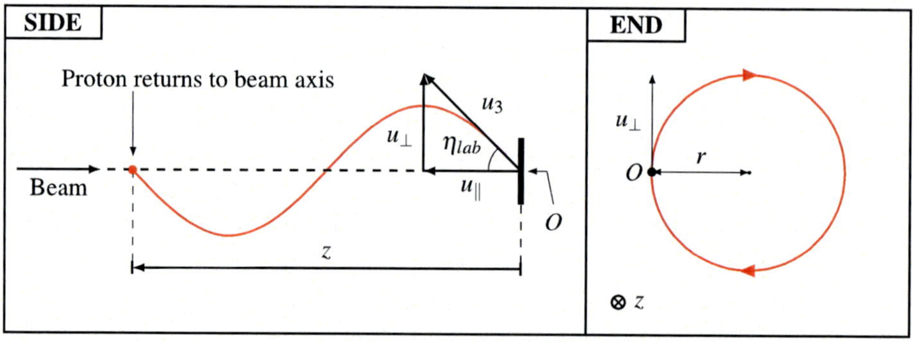

Fig. 4.11 A diagram showing the path of an ejectile proton within the solenoid from a (d,p) reaction. The distance along the beam axis z is negative here, because it is in the opposite direction to the beam

Table 4.1 Orbital time periods of light particles in a solenoid with a magnetic field of 2.5 T

Particle	q	m_3 (u)	t_{cyc} (ns)
p	1	1.0073	26.2
d	1	2.0136	52.4
t	1	3.0155	78.5
^{3}He	2	3.0149	39.3
α	2	4.0015	52.1

As the ejectiles are scattering with a component of velocity antiparallel to the beam direction, the angles θ_{lab} and $\theta_{cm} > \frac{\pi}{2}$, they will be relabelled with $\eta_{lab} = \pi - \theta_{lab}$ and $\eta_{cm} = \pi - \theta_{cm}$ for convenience.

Using Eq. 4.13, the time period for this orbit, t_{cyc} is therefore

$$t_{cyc} = \frac{2\pi r}{u_\perp} = \frac{2\pi m_3}{qB} = \frac{2\pi}{\omega}, \tag{4.14}$$

where ω is the cyclotron period for the ejectile. This result is important—knowledge of the magnetic field strength and a measurement of the time period is sufficient to give the mass-to-charge ratio of a given ejectile, providing a means for identifying particles. For light ejectiles, these ratios are sufficiently different that nanosecond-timing resolution allows distinctions between most particles to be made, as shown in Table 4.1.

Where particles have similar mass-to-charge ratios, they can be typically identified by other measurements, such as the amount of energy deposited in the detector. Hence, there is much easier ejectile identification for reactions in IK in ISS than using the deposited energy, which, as described in Sect. 4.3, is small for reactions in IK.

The distance from the target along the beam line, z, is related to t_{cyc} by the parallel component of the ejectile velocity:

$$z = u_\parallel t_{cyc}. \tag{4.15}$$

If the ejectile completes multiple orbits, then z can simply be scaled by this number to calculate where the ejectile returns to the axis. As protons tend to be emitted backwards in the lab frame ($\eta_{lab} < \frac{\pi}{2}$) for a (*d*,*p*) reaction, z is negative. The two components of velocity in the lab frame can be related to the corresponding velocities in the CM frame using Fig. 4.7,

$$u_\parallel = u_3 \cos\theta_{lab} = v_{cm} + v_3 \cos\theta_{cm} = v_{cm} - v_3 \cos\eta_{cm}, \tag{4.16}$$

$$u_\perp = u_3 \sin\theta_{lab} = v_3 \sin\theta_{cm} = v_3 \sin\eta_{cm}. \tag{4.17}$$

Equation 4.7 can then be rewritten in terms of z using Eqs. 4.14–4.16 to give

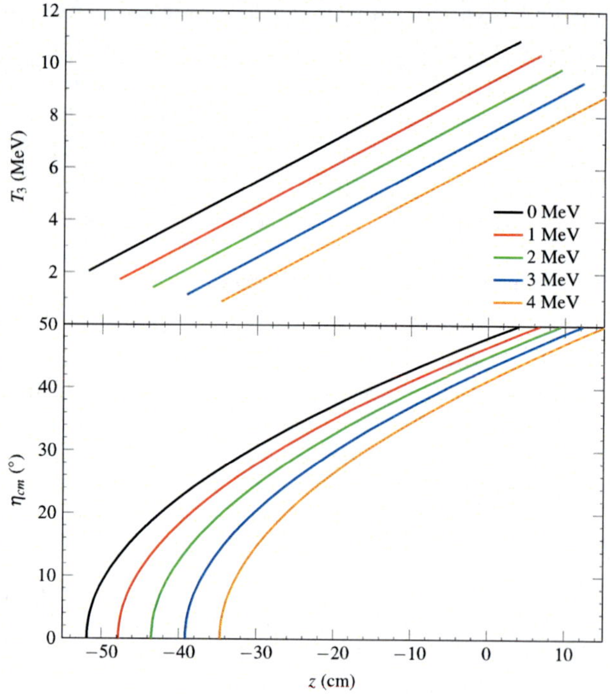

Fig. 4.12 Kinematic lines for ejectile protons produced in the $d(^{28}\text{Mg},p)^{29}\text{Mg}$ reaction in the range $0° \leqslant \eta_{cm} \leqslant 50°$. This assumes that the protons return to the beam axis. The top plot shows T_3 against z for a singular orbit. The bottom plot shows the mapping of η_{cm} onto z

$$T_3 = T_3^{cm} - \frac{1}{2} m_3 v_{cm}^2 + \frac{qBv_{cm}}{2\pi} z. \tag{4.18}$$

The first two terms in this equation are constants for a given excitation energy, so the ejectile energy is linearly dependent on z. This is in contrast to Eq. 4.7, where the ejectile energy is linearly dependent on $\cos\theta_{cm}$, so non-linear in θ_{cm}. Therefore, a conventional fixed-angle experiment will have a non-linear energy dependence on the angle—the cause of the unwanted kinematic effects of KC and KS. The solenoid technique avoids these kinematic effects, because the ejectile energy is linearly proportional to z. The angles and excitation energy can be extracted from the measurement of both T_3 and z.[4] This relationship is demonstrated graphically in Fig. 4.12.

It can be clearly seen that the effect of KC has been eliminated, as the difference in T_3 is proportional to the difference in E_x for a given z. This difference in T_3 is solely due to the kinetic energy of the ejectile in the CM frame, T_3^{cm}, as v_{cm} does not vary with E_x.

While the analogue of KC is absent in the solenoid, there is an analogue of KS because $\frac{dT_3}{dz} \neq 0$. The magnitude of the KS in each technique can be directly compared. Using Eq. 4.9, the KS for a conventional fixed-angle experiment, where θ_{lab} is measured, is

[4] This is demonstrated in Appendix A.

$$\frac{\mathrm{d}T_3}{\mathrm{d}\theta_{lab}}\Delta\theta_{lab} = -m_3 v_3 v_{cm} \sin\theta_{cm} \frac{\mathrm{d}\theta_{cm}}{\mathrm{d}\theta_{lab}}\Delta\theta_{lab}. \tag{4.19}$$

Using Eq. 4.18, the analogue KS for the solenoid technique, where z is measured, is

$$\frac{\mathrm{d}T_3}{\mathrm{d}z}\Delta z = \frac{qBv_{cm}}{2\pi}\Delta z. \tag{4.20}$$

For a fixed-angle technique, the angular resolution achievable is on the order of a few degrees to be able to obtain a significant yield. For example, the TREX experiment at ISOLDE, which detects ejectiles from reactions in IK with a fixed-angle set-up, has an angular resolution of 2°–6° in θ_{lab} [32]. For the solenoid, the position on a detector can be known on the order of a few millimetres.

The contributions to the resolution from each of these techniques can then be directly compared by taking the ratio of Eq. 4.19–4.20,

$$\left|\frac{\frac{\mathrm{d}T_3}{\mathrm{d}\theta_{lab}}\Delta\theta_{lab}}{\frac{\mathrm{d}T_3}{\mathrm{d}z}\Delta z}\right| = \frac{2\pi m_3 v_3 \sin\theta_{cm}}{qB\Delta z}\frac{\mathrm{d}\theta_{cm}}{\mathrm{d}\theta_{lab}}\Delta\theta_{lab}. \tag{4.21}$$

Values of this ratio are listed in Table 4.2 for different values of η_{cm} for protons emitted where the ground state of ^{29}Mg has been populated.

This shows that the contribution to the resolution from KS is much larger for conventional fixed-angle setups than for solenoidal spectrometers. The only case where they do better is when the position resolution in the solenoid is of the order of 1 cm or more at very low angles. However, positions in the solenoid can be trivially measured on the order of millimetres, showing the superiority of the solenoid technique in this contribution to the energy resolution.

The solenoid technique additionally benefits from efficient solid angle coverage for charged particles, due to the focusing effects of the magnetic field. Instead of surrounding the target with detectors, the detectors can be placed around the beam line. However, the introduction of a detector with a non-zero radius on the beam line affects the z-dependence of measured and deduced properties, and this will be discussed below.

Table 4.2 Values of the fraction in Eq. 4.21 for different values of $\Delta\theta_{lab}$, Δz, and η_{cm}

Δz (cm)	(for $\Delta\theta_{lab} = 2°$)			(for $\Delta\theta_{lab} = 6°$)		
η_{cm} (°)	0.01	0.1	1.0	0.01	0.1	1.0
5	16.4	1.6	0.2	49.2	4.9	0.5
10	36.2	3.6	0.4	108.5	10.8	1.1
15	62.0	6.2	0.6	186.1	18.6	1.9
20	95.8	9.6	1.0	287.3	28.7	2.9
25	137.8	13.8	1.4	413.3	41.3	4.1
30	187.5	18.8	1.9	562.6	56.3	5.6

4.4.2 Kinematics for a Detector of Finite Size

Corrections to the kinematics calculations are necessary due to the finite size of the detector. In reality, ejectiles do not return to the beam axis, but are incident on the detector before reaching it. A diagram of this is shown in Fig. 4.13.

Instead of completing one whole orbit, the ejectile moves helically through an angle $2\pi - \alpha$. This angle can be rewritten as

$$\alpha = 2 \arcsin\left(\frac{\rho}{2r}\right), \tag{4.22}$$

where ρ is the radius of the detector, equal to the distance OP. When considering the full ϕ coverage within the detector, the radius of the detector, ρ, will change, as the cross section of the detector is a square. To take account of this, an average detector radius, ρ_0 can be calculated and used in place of ρ. This is illustrated in Fig. 4.14.

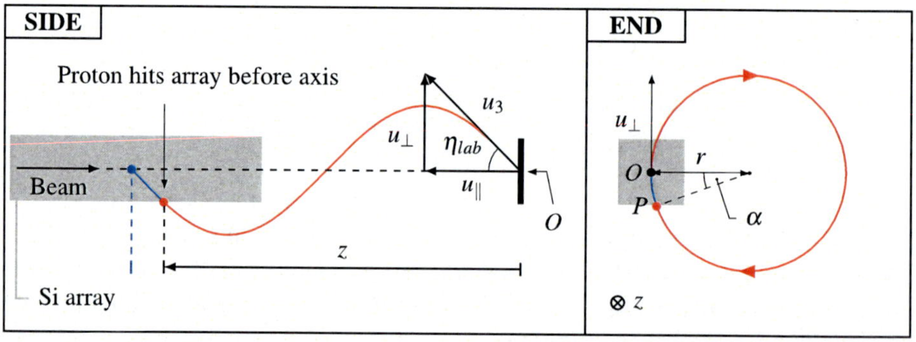

Fig. 4.13 A diagram showing the effects on the trajectory of an ejectile when a finite detector is placed on the beam axis. The blue path indicates where the ejectile would have travelled

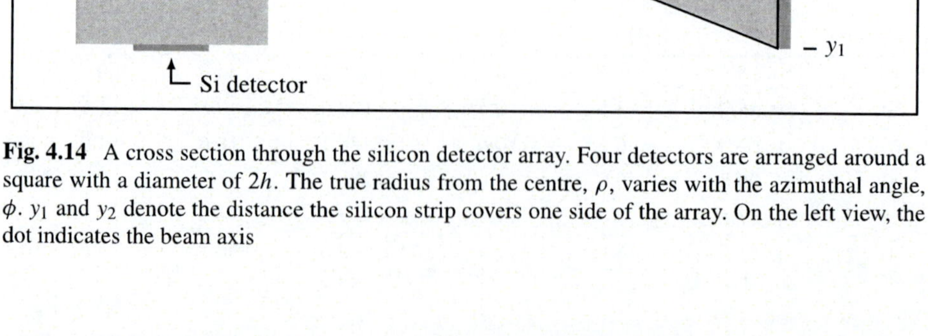

Fig. 4.14 A cross section through the silicon detector array. Four detectors are arranged around a square with a diameter of $2h$. The true radius from the centre, ρ, varies with the azimuthal angle, ϕ. y_1 and y_2 denote the distance the silicon strip covers one side of the array. On the left view, the dot indicates the beam axis

The average radius, ρ_0, is therefore

$$\rho_0 = \frac{\displaystyle\int_{y_1}^{y_2} \rho(y)\,\mathrm{d}y}{\displaystyle\int_{y_1}^{y_2} \mathrm{d}y} = \frac{\displaystyle\int_{y_1}^{y_2} \sqrt{h^2 + y^2}\,\mathrm{d}y}{(y_2 - y_1)}. \tag{4.23}$$

Using the substitution $y = h\tan\phi$, ρ_0 becomes

$$\rho_0 = \frac{1}{2(y_2 - y_1)} \left\{ y_2\sqrt{h^2 + y_2^2} - y_1\sqrt{h^2 + y_1^2} + h^2 \ln\left| \frac{y_2 + \sqrt{h^2 + y_2^2}}{y_1 + \sqrt{h^2 + y_1^2}} \right| \right\}. \tag{4.24}$$

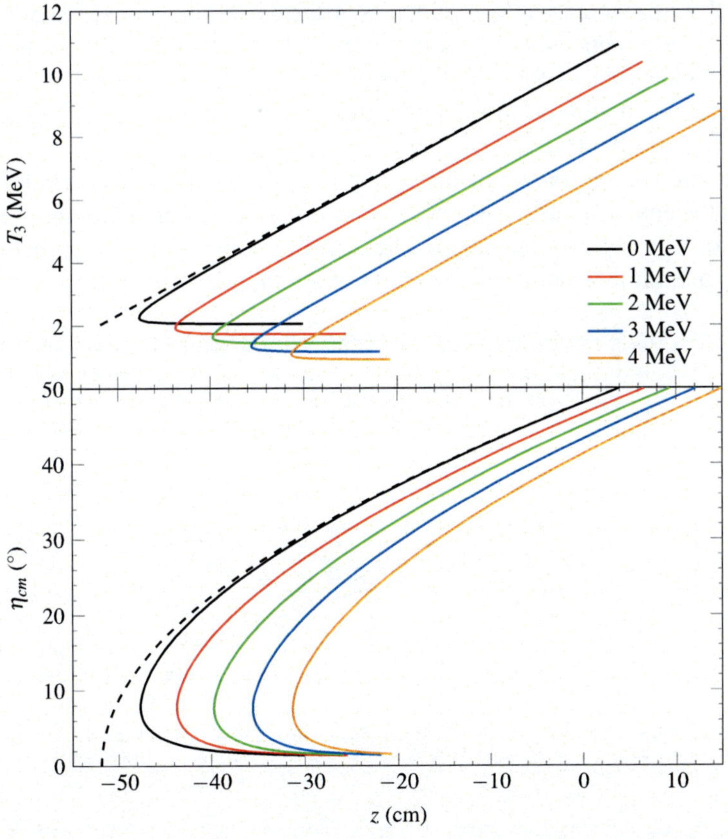

Fig. 4.15 Kinematic lines for the $d(^{28}\mathrm{Mg},p)^{29}\mathrm{Mg}$ reaction for a detector of a finite size in the angular range $0° \leqslant \eta_{cm} \leqslant 50°$. The top plot shows T_3 against z for a singular orbit. The bottom plot shows the mapping of η_{cm} onto z. The dimensions of this detector are detailed in Sect. 4.6. The dashed line is the kinematic line for the ground state where the ejectile returns to the beam axis

As each side of the silicon detector array is identical, this average radius holds for the whole array.

The time to complete one orbit must be modified to include the angle α, so Eq. 4.14 becomes

$$t_{cyc} = \frac{r(2\pi - \alpha)}{u_\perp} = \frac{2m_3}{qB}\left[\pi - \arcsin\left(\frac{\rho_0}{2r}\right)\right]. \tag{4.25}$$

This transforms z which becomes

$$z = \frac{2m_3 u_\parallel}{qB}\left[\pi - \arcsin\left(\frac{\rho_0 qB}{2m_3 u_\perp}\right)\right], \tag{4.26}$$

which is plotted against T_3 and η_{cm} in Fig. 4.15.

The kinematic lines here are no longer single-valued and linear, but instead have "knees" at low values of η_{cm}, which causes double-valued portions of the function. The knees occur as the difference between the value of z in the simple case and the actual value of z increases. As η_{cm} increases, the deviation of the kinematic lines from the simple case reduces. If the radius of the detector array increases, the knees will occur at higher values of η_{cm}, reducing the useful angular coverage for the detector. It is, therefore, important to minimise this radius while still allowing the beam to pass through the centre.

While an average radius of the silicon detector array has been calculated, this replaces the actual square cross section with an idealised circular cross section. The validity of this assumption can be calculated by looking at the difference in z for particles that hit the centre and the edges of a particular silicon strip on the detector

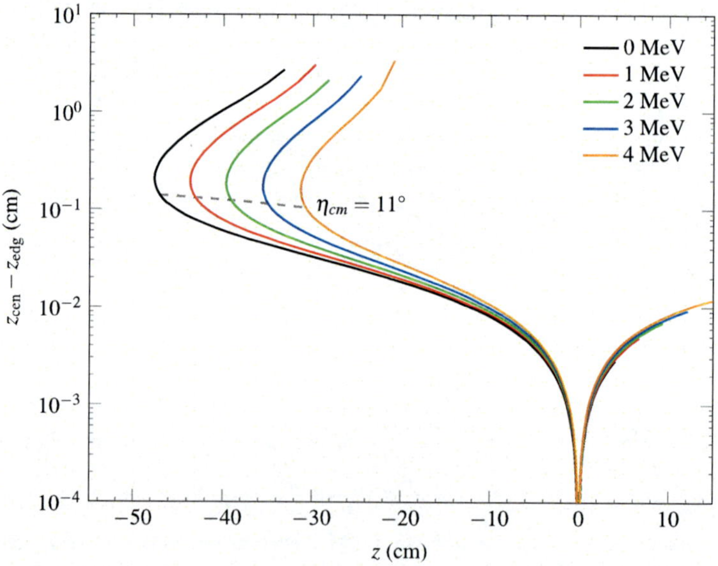

Fig. 4.16 The difference in z for particles hitting a given silicon detector in the middle or on the edge, for the value of z when using the average radius ρ_0 for the square array used. These values are plotted for the range $0° \leqslant \eta_{cm} \leqslant 50°$. The "knees" appear above the dashed line, for $\eta_{cm} < 11°$

array. The radius of the array will vary between two values, ρ_{cen} and ρ_{edg}. Ejectiles hitting the silicon strip at these radii for a given θ_{cm} will have different values of z, labelled z_{cen} and z_{edg}. The difference this makes is illustrated in Fig. 4.16.

Figure 4.16 shows the difference in z between z_{cen} and z_{edg}, the values of z calculated at the minimum and maximum detector radius respectively. After the knees, the difference this makes to the position is of the order 0.1–0.01 cm for the majority of values of z, well within the total position resolution of the detectors. Therefore, assuming that the radius of the array remains constant is reasonable.

4.5 The Solenoid

An Oxford OR66 solenoid was used to provide the magnetic field for ISS. It is a repurposed MRI magnet that is able to sustain fields up to 4 T. The bore of the magnet is 925 mm, and the detectors within the magnet are aligned to this magnetic-field axis down the centre. The diameter of the total magnet is 2396(10) mm, and has a length of 2732(5) mm [27]. The radial field, B_ρ, and the axial field, B_z, for ISS are shown in Fig. 4.17. For the $d(^{28}\text{Mg},p)^{29}\text{Mg}$ experiment, the axial field was set to 2.5 T.

Up to this point, the kinematics of particles in ISS have assumed a completely uniform magnetic field. In reality, this is an idealised case, as Fig. 4.17 clearly shows. Trajectories at higher ρ or at higher z will have both a changing axial field and radial field. Results from a GEANT4 [33] simulation indicate that a 1% change in axial field can increase the FWHM of a peak by approximately 10% [34]. This will primarily affect ejectiles with high-η_{cm} because they are more likely to reach larger ρ in the solenoid. It can also affect ejectiles with low η_{cm} if they reach a large value of z where the field is inhomogeneous, but this is less likely.

Therefore, confining the ejectiles to the uniform region of the magnet is crucial for optimising the position resolution on the array. This region can be defined by the maximum values of ρ and z for all desired particle trajectories. Using Eqs. 4.13 and 4.17, the maximum radius within the detector occurs at the diameter of the helical orbit:

$$\rho_{\max} = 2r = \frac{2m_3 u_\perp}{qB} = \frac{2m_3 v_3 \sin\eta_{cm}}{qB}. \tag{4.27}$$

$\rho_{\max}$ can be maximised by using the maximum η_{cm} of interest for protons, which is 40°, and having $E_x = 0$ to maximise v_3. This yields $\rho_{\max} = 32.0$ cm. This is not surpassed by the equivalent for deuterons, which is $\rho_{\max} = 25.5$ cm for $\eta_{cm} = 22.491°$.

The maximum z for this region occurs when the proton scatters backwards at $\eta_{cm} = 0$ and completes one full "orbit". Using Eqs. 4.14–4.16, the maximum distance travelled from the target is

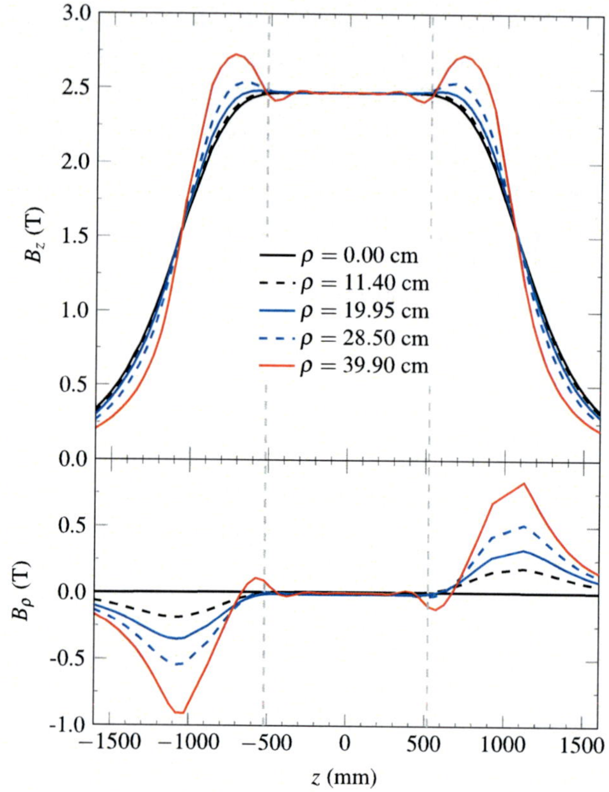

Fig. 4.17 A map of the field strength measured a few months prior to the $d(^{28}Mg,p)^{29}Mg$ experiment taking place, where the field was 2.47 T on the beam axis in the centre of the solenoid. The grey dashed lines are at ±520 mm and correspond to z_{max} (defined in Eq. 4.28) for $E_x = 0$ at $\eta_{cm} = 0°$

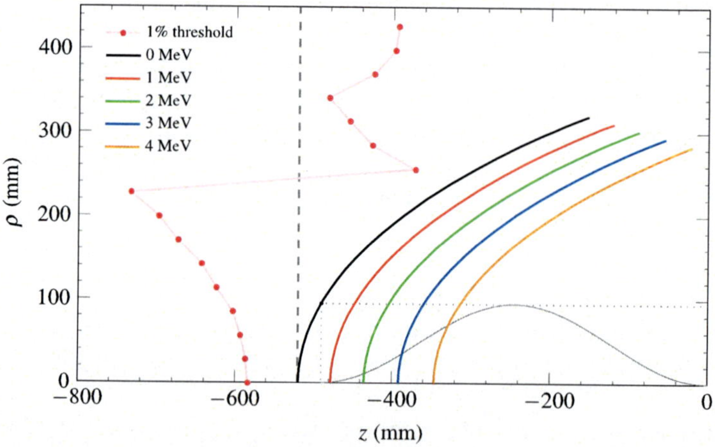

Fig. 4.18 The variation of ρ_{max} and z_{max} for the angular range $0° \leqslant \eta_{cm} \leqslant 40°$. The axial field 1% contour has been drawn for comparison. The grey dashed line is at $z = -520$ mm. The grey dotted line shows the boundaries of the trajectory for an ejectile that populated the ground state of ^{29}Mg. This trajectory has been drawn within these boundaries, with ϕ chosen so that the initial $u_\perp$ is into the plane of the paper

$$z_{\max} = u_{\parallel} t_{cyc} = \frac{2\pi m_3}{qB}(v_{cm} - v_3 \cos \eta_{cm}) = \frac{2\pi m_3}{qB}(v_{cm} - v_3) \tag{4.28}$$

For a 2.5 T axial field, this corresponds to −0.52 m, which has been marked on Fig. 4.17. This can be more helpfully visualised by plotting the variation of $\rho_{\max}$ and $z_{\max}$ with η_{cm} with the 1% field variation threshold marked. This is shown in Fig. 4.18.

The lines corresponding to $\rho_{\max}$ and $z_{\max}$ can be used to work out the boundaries of the trajectory of ejectiles in the solenoid for the given excitation energy, as demonstrated for a particular ejectile in Fig. 4.18. Each trajectory is within a cylinder of radius $\rho_{\max}$ and length $z_{\max}$ a distance r from the z-axis. All of these possible trajectories are within the 1% threshold, showing that it is valid to treat the field as constant for this experiment.

4.6 Detectors in the ISS

A diagram of the detectors in ISS for the $d(^{28}\mathrm{Mg},p)^{29}\mathrm{Mg}$ reaction is shown in Fig. 4.10, where the paths of protons, deuterons, and recoiling nuclei are highlighted. The detectors used to detect these particles are discussed here.

4.6.1 The HELIOS Silicon Array

As previously mentioned, the HELIOS silicon detector array was used to detect the ejectile protons. Photographs of the array and a PSD are shown in Figs. 4.19 and 4.20.

The array contains 24 silicon position-sensitive detectors (PSDs). These are mounted on a square extruded aluminium rod to form an array of detectors with six detectors on each side. The dimensions for both the rod and the PSDs are in Table 4.3.

The rod dimensions were chosen to allow the beam to pass through the centre of the rod and also retain the desired kinematic properties as outlined in Sect. 4.4, namely that the radius of the detector array is minimised to reduce the size of the "knees" on the energy-position spectrum. An adjustable four-jaw slit was attached to the end of the array nearest the target, but was not used in this experiment.

Each PSD produces three signals: a total energy signal from an aluminium contact on the back of the detector, E, and two signals at each end of the strip from aluminium contacts at the front of the detector, X_1 and X_2. This is illustrated in Fig. 4.21.

There is approximately 17 kΩ resistance between X_1 and X_2, and this resistance allows the position to be calculated using resistive charge division. The charge collected at X_1 and X_2 will be inversely proportional to the distance from the interaction site in the PSD.

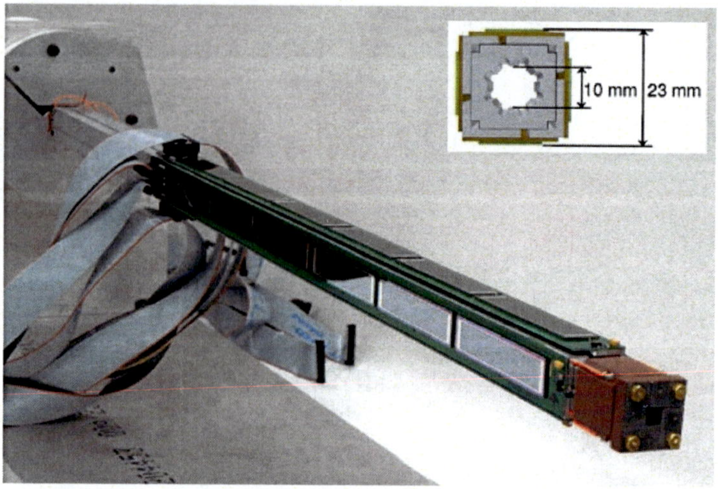

Fig. 4.19 A photograph of the HELIOS detector array [35]. The four-jaw slit can be seen in red at the end of the array

Fig. 4.20 A photograph of a PSD mounted to a PC board [35]

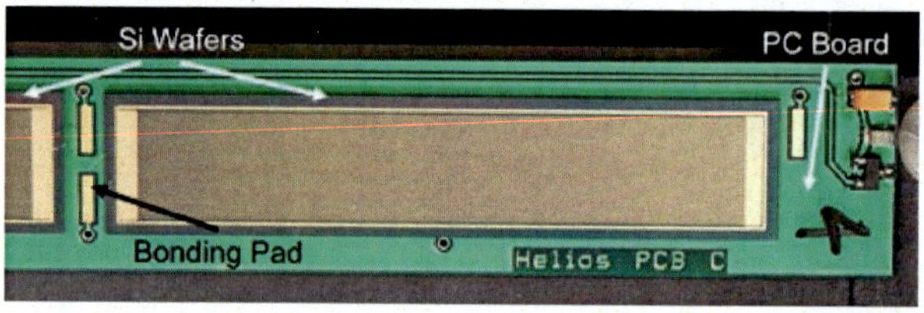

Table 4.3 Measurements of the PSDs and rod used in ISS, taken from Refs. [35, 36]

PSDs		Rod	
Dimensions	12 mm × 56 mm	Outer diameter	23 mm
Thickness	700 ± 15 μm	Inner diameter	10 mm
Active area	9 mm × 50.5 mm	Total length	710 mm
Array active area	9 mm × 345.3 mm[a]		

[a] This length represents the distance from the beginning of the first PSD to the end of the last PSD on a given side. However, there are gaps between each PSD which will not detect any particles. This active area is the area which the detectors occupy on the array including these gaps. The difference between this length and the total length of 6 detectors with no gaps is 42.3 mm

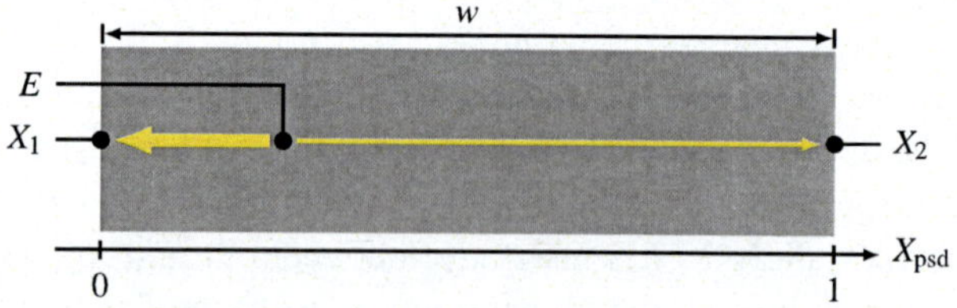

Fig. 4.21 A diagram of the PSD. A charge impacts on the detector giving a total energy signal E. Signals are additionally collected at X_1 and X_2 that can be used to infer the position on the strip. The X_{psd} scale is defined in Eq. 4.29

The position on the strip can then be determined using

$$X_{\rm psd} = \begin{cases} 1 - \dfrac{X_1}{E}, & X_1 > \dfrac{E}{2} \\ \dfrac{X_2}{E}, & X_2 \geqslant \dfrac{E}{2} \end{cases}, \tag{4.29}$$

where $0 \leqslant X_{\rm psd} \leqslant 1$, and has been labelled in Fig. 4.21. This can be scaled to the width of the PSD, w, and thus used to calculate the distance from the target to the position on the strip. If X_2 is closer to the target than X_1, then

$$z = w\left(X_{\rm psd} - \frac{1}{2}\right) - d_{ta} - d_i, \tag{4.30}$$

where d_{ta} is the distance from the target to the array, and d_i is the distance from the end of the array to the centre of detector i. The positions of the detectors on the array are listed in Table 7.2 of Ref. [36].

The position resolution has been previously established in Ref. [36] using elastically scattered protons from a ^{12}C foil at 2.0, 3.0, 4.0, and 5.0 MeV. A mask with eight 0.5 mm slits was used to cover the PSD, and the average FWHM of the resultant peaks was extracted. This was 0.532 mm at 5.0 MeV and 1.17 mm at 2.0 MeV. The energy resolution was also measured in this scattering, and was found to be 27–53 keV at FWHM, depending on the detector.

This energy resolution includes a contribution from the position resolution, due to the energy dependence on z. This was characterised in Eq. 4.20, but can be more helpfully represented as

$$\frac{{\rm d}T_3}{{\rm d}z} = \frac{1}{\pi\sqrt{2}} qB \frac{m_1}{m_1 + m_2}\sqrt{\frac{T_1}{m_1}} \approx 2.21 qB \frac{m_1}{m_1 + m_2}\sqrt{\frac{T_1}{m_1}}\left[\frac{\rm keV}{\rm mm}\right], \tag{4.31}$$

where $\frac{T_1}{m_1}$ is the beam energy in MeV, q is the charge in units of the elementary charge, and B is the magnetic field in units of Tesla. For the $d(^{28}\text{Mg},p)^{29}\text{Mg}$ reaction, which had a field of 2.5 T and a beam energy of 9.473 MeV/u, then $\frac{{\rm d}T_3}{{\rm d}z} \approx 16.4$ keV/mm. This contribution to the energy resolution will be discussed further with other contributions in Sect. 5.1.3.

The solid angle for the detectors can be seen by examining the trajectories of the ejectile protons—examples for the $d(^{28}\text{Mg},p)^{29}\text{Mg}$ experiment are shown in Fig. 4.22. The azimuthal angular coverage for the array can be seen from the trajectories in Fig. 4.22. It can be calculated due to the symmetry of the detectors around the beam axis for the length of the array. Using the set-up in Fig. 4.14, the azimuthal angular coverage for one detector is

$$\Delta\phi = 2\arctan\left(\frac{y_2}{h}\right), \tag{4.32}$$

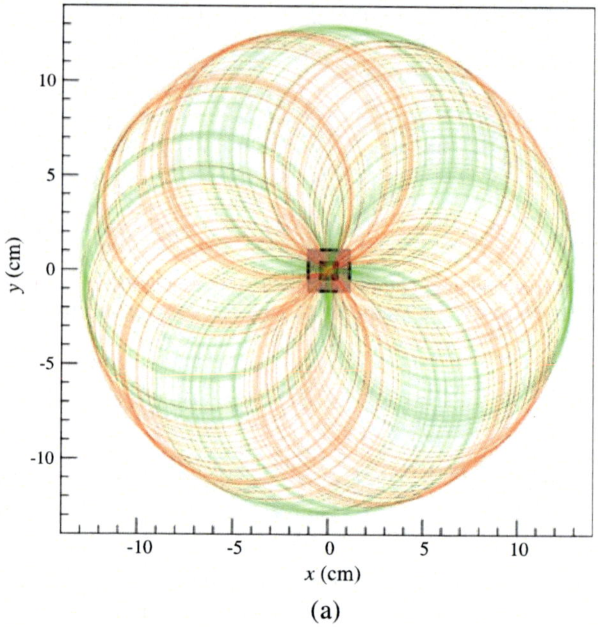

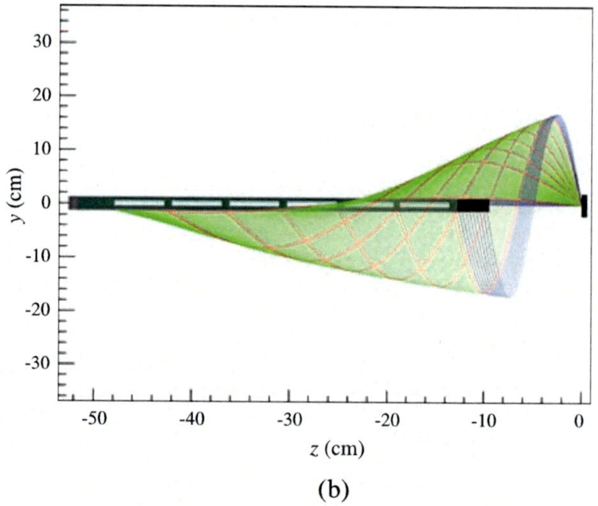

Fig. 4.22 Trajectories of protons on the array with $E_x = 0$ MeV. Green trajectories hit the PSDs, whereas red trajectories hit the array frame. Blue trajectories are those that hit the four-jaws before the array. **a** is viewed from the target towards the array for $\eta_{cm} = 15°$ for pseudorandom values of ϕ. **b** is viewed from the side for $0° \leqslant \eta_{cm} \leqslant 43.2°$ (single-orbit trajectories). The spacing between these trajectories is 0.2°. They have their initial values of $u_\perp$ aligned in the positive y-direction

where $y_2 = -y_1$. Using the width of the active area and the size of the outer diameter from Table 4.3, $\Delta\phi \approx 0.746$ radians, which is approximately 42.7° per detector. Therefore, the total efficiency in ϕ is 47%, which agrees with what is seen in Fig. 4.22, where roughly half of the trajectories hit a detector.

The polar solid angle coverage is represented in Fig. 4.22, which shows all single-orbit trajectories that hit the detector. This polar coverage can be calculated and combined with $\Delta\phi$ from Eq. 4.32 to place an upper limit on the total solid angle of the detector array. Using Eqs. 4.15 and 4.16, the total solid angle for ejectiles that

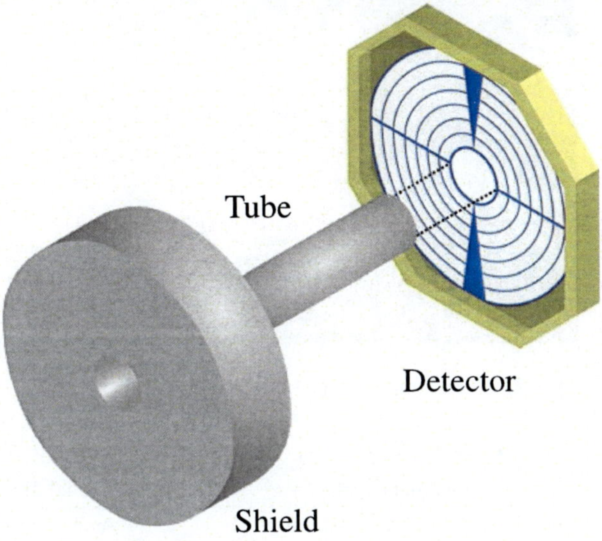

Fig. 4.23 Schematic of the luminosity detector (not to scale)

carry out single turns before hitting the array is

$$\Delta\Omega = \int \mathrm{d}\phi \int \sin\theta_{cm}\mathrm{d}\theta_{cm} = -4\Delta\phi \int \frac{\mathrm{d}z}{v_3 t_{cyc}} = \frac{4\Delta\phi\Delta z}{v_3 t_{cyc}}, \quad (4.33)$$

where Δz is the total length that the detectors occupy in the z-direction. The total solid angle, $\Delta\Omega$, is therefore independent of the position of the detectors within the solenoid, though there is a limit to the value of Δz from the angle limit imposed by the position of the target in the solenoid. For the ground state, and using a total length of 303 mm from the 6 detectors on each side of the array, $\Delta\Omega = 0.58$ sr, or just under 5% of 4π. This is an upper limit, however, as cuts were applied to the detectors, and this is detailed more in Sect. 5.4.2.

4.6.2 The Luminosity Detector

The luminosity detector is the first detector that is downstream from the target, as shown in Fig. 4.10. It detects elastically scattered deuterons in the reaction. A schematic of the detector setup is shown in Fig. 4.23, and some photographs are shown in Fig. 4.24.

The detector itself is a Micron Semiconductor S1 detector [37], a double-sided silicon detector. It is annular in shape, and shielded by both a thick aluminium annulus and a tube that passes through the centre of the annulus. Therefore, particles can only enter the detector by passing over the shield. The shield and tube prevent many unwanted particles from hitting the detector, but also reduces the solid angle for

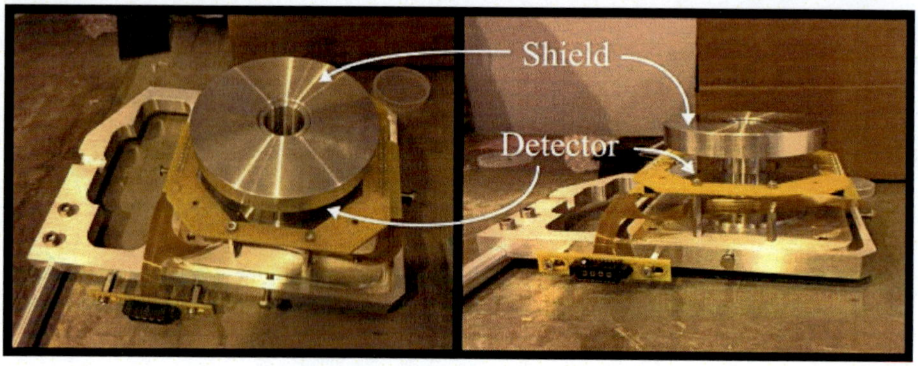

Fig. 4.24 Photographs of the luminosity detector

Table 4.4 Relevant dimensions of the luminosity detector within ISS

Distances along the beam axis		Radial distances	
Thickness of Al shield	12.8 mm	Inner diameter	48.00 mm
Distance between detector and shield	13.42 mm	Outer diameter	96.00 mm
Target-S1 distance	125.7 mm		

deuterons to hit it. The distances for the luminosity detector within ISS are presented in Table 4.4.

Because of the shielding, there are only two trajectories that the deuterons can take before hitting the detector, corresponding to one and two revolutions in a helical orbit, and these are presented in Fig. 4.25. These can be distinguished, as they have different energies according to Eq. 4.7. With increased revolutions comes a reduced solid angle, so the single-revolution peak was used to evaluate the target thickness.

The detector is divided into four quadrants that give full 2π coverage in ϕ for any scattered deuterons that can enter the detector.

4.6.3 *Beam Diagnostics Detectors*

Two beam diagnostics devices were placed downstream from the luminosity detector, as was shown in Fig. 4.10. Photographs of these devices from within ISS are in Fig. 4.26. These were only used for examining the beam without any obstruction from a target; otherwise they were moved out of the path of the beam. These devices were a Faraday cup, for monitoring the beam current, and a ΔE–E detector to test the purity of the beam. The ΔE–E detector was composed of two ORTEC silicon surface barrier detectors,[5] with an active area of 150 mm^2 [38].

[5] The model numbers were B-018-150-500 and D-030-150-75 respectively.

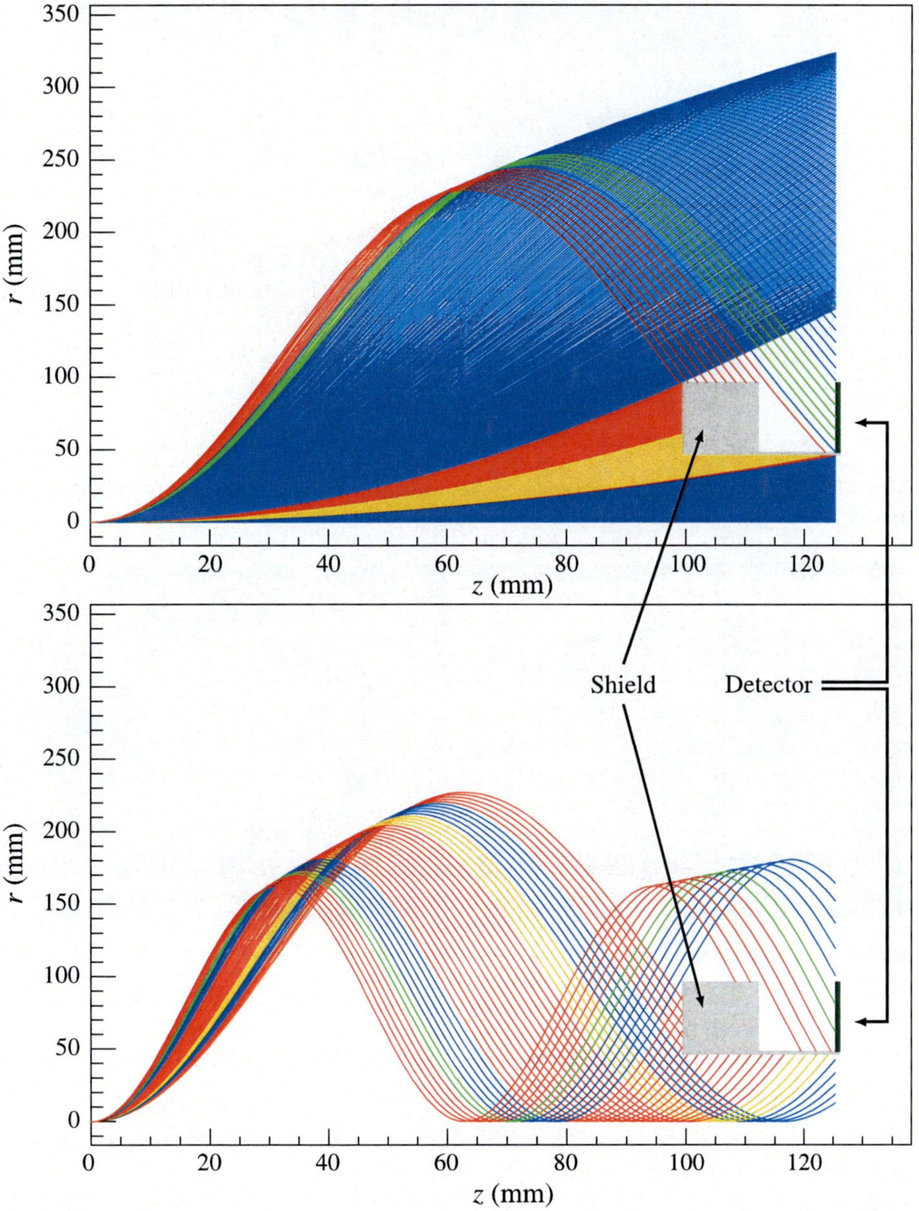

Fig. 4.25 Trajectories of deuterons in the ISS. The initial component of deuteron velocity perpendicular to the beam line is directed into the plane of the paper. Most trajectories miss both the shield and the detector (blue), some hit the shield which would have hit the detector (orange), and some are just blocked by the shield (red). Only a fraction of the trajectories hit the detector (green). The spacing between trajectories here is 0.15°, and the angular range for successful trajectories is $21.615° \leqslant \theta_{cm} \leqslant 22.491°$ for single-orbit trajectories (top), and $14.755° \leqslant \theta_{cm} \leqslant 15.127°$ for double-orbit trajectories (bottom)

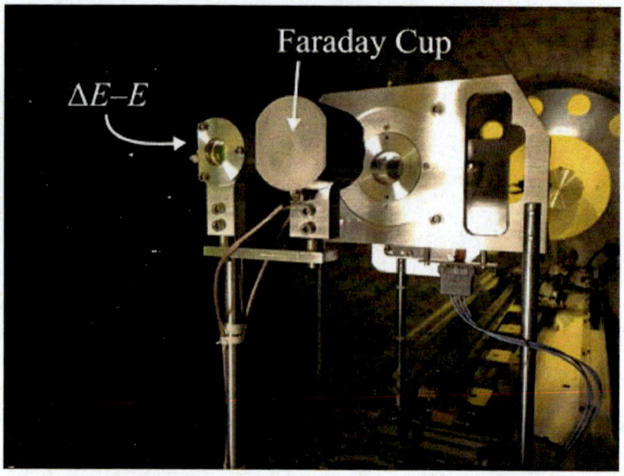

Fig. 4.26 The 0° ΔE–E detector (left) and the Faraday cup (right) behind the luminosity detector in ISS [27]

Fig. 4.27 Photographs of the recoil detector

4.6.4 The Recoil Detector

The recoil detector was the final detector used downstream from the target, as was shown in Fig. 4.10. Photographs of the assembled recoil detector are shown in Fig. 4.27.

The detector consists of eight Micron QQQ1 detectors [39], which are single-sided silicon detectors. They are arranged in two layers of four detectors each to form a ΔE–E detector. They were used to detect the recoiling residual nuclei from the reaction. Any unreacted beam was able to pass through the centre into a beam dump. The dimensions of each of these detectors is detailed in Table 4.5.

Although there is not complete 2π-coverage in ϕ, the residual nuclei of interest are correlated via two-body kinematics with the proton detection on the four sides of the array. Therefore, the recoil array was rotated in order to fully cover the recoils

Table 4.5 Dimensions of the individual QQQ1 detectors used in the recoil detectors [39]

Radii (mm)		Thickness and angular coverage	
Chip inner	11.50	Angular coverage (°)	82
Chip outer	103.00	ΔE thickness (μm)	47, 51, 64, 71
Si inner	18.00	E thickness (μm)	494, 498, 501, 502
Si outer	100.00		

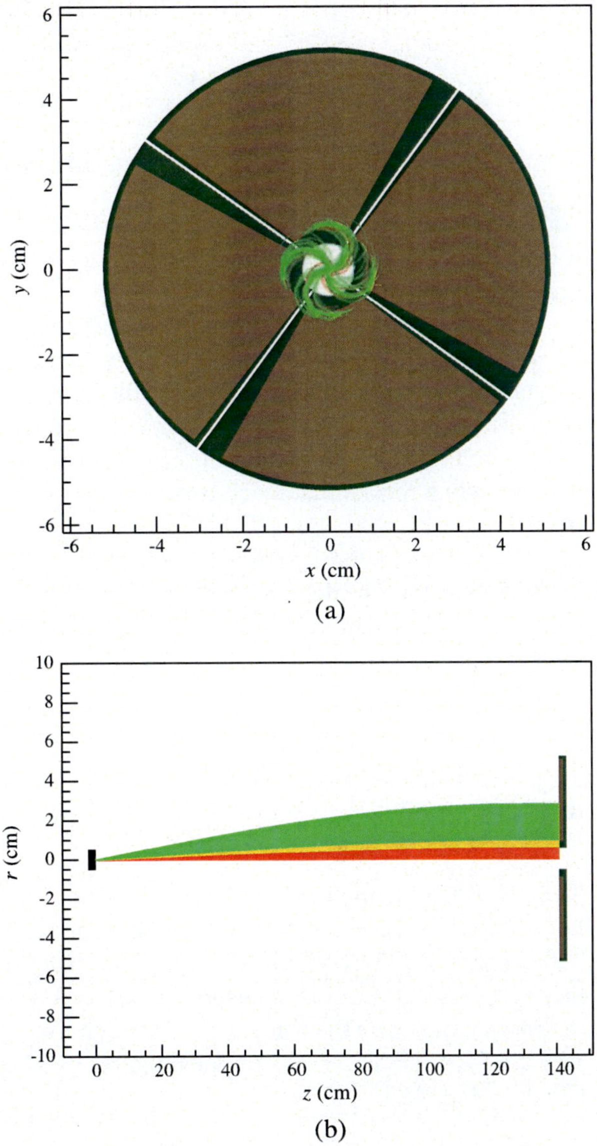

Fig. 4.28 Recoil trajectories from the $d(^{28}Mg,p)^{29}Mg$ reaction. Trajectories in green hit the recoil detector, whereas those in yellow hit the surrounding material, and those in red pass through the middle. Successful trajectories occur when $\eta_{cm} > 13.6°$. **a** Viewpoint from the target for $\eta_{cm} = 15°$. **b** Viewpoint from the side for differing η_{cm}

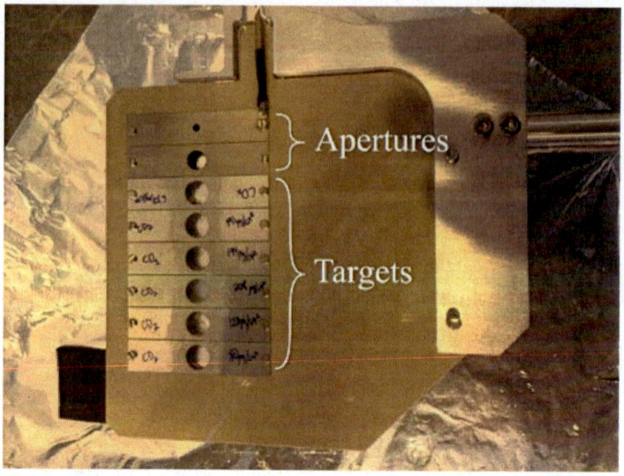

Fig. 4.29 The target ladder used in ISS. 6 targets were mounted, and there were additional apertures mounted for beam tuning [27]

correlated with detected protons in the array. Using the target–recoil detector distance, this angle was calculated to be 36° from the vertical. The resultant trajectories of recoiling nuclei are shown in Fig. 4.28.

These trajectories fall into a windmill-pattern, corresponding to the four sides of the array. Like the ejectiles, the residual nuclei move in helical orbits, but have a longer orbital time period than the ejectiles because of their larger mass-to-charge ratio. They hit the recoil detector partway through their helical orbit. While only single orbits have been displayed, ejectiles that undergo multiple turns will have corresponding recoils that reach further distances from the beam axis on the recoil detector.

The recoils pass through the ΔE detector and deposit some energy and then stop in the E detector. The deposition of energy in both of these detectors is described by the Bethe-Bloch formula [40], and different masses and energies can be distinguished by their energy-loss characteristics. Thus the recoil detector can be used to identify the residual nuclei of interest.

4.7 Targets

The targets used in this experiment were a series of CD_2 targets with a range of thicknesses. The main analysis was carried out using targets of thickness 80 μg cm^{-2} and 120 μg cm^{-2} as these provided sufficiently many reactions without compromising the energy resolution. This corresponded to two different positions of the silicon array, in order to maximise the solid angle coverage. An additional target of 637 μg cm^{-2} was used for validation of array positioning. A picture of the target ladder used is shown in Fig. 4.29.

References

1. Catherall R, Andreazza W, Breitenfeldt M, Dorsival A, Focker GJ, Gharsa TP, Giles TJ, Grenard JL, Locci F, Martins P et al (2017) The ISOLDE facility. J Phys G: Nucl Part Phys 44(9):094002
2. Reich KH (1969) The CERN proton synchrotron booster. CERN-SI-Int-DL-69-1. CERN-CDS: 349912
3. Hanke K (2013) Past and present operation of the CERN PS booster. Int J Mod Phys A 28(13):1330019
4. Bjørnstad T, Hagebø E, Hoff P, Jonsson OC, Kugler E, Ravn HL, Sundell S, Vosicki B, Collaboration The ISOLDE (1986) Methods for production of intense beams of unstable nuclei: new developments at ISOLDE. Phys Scr 34(6A):578–590
5. Rodriguez JA (2018) 2018 REX/HIE-ISOLDE Physics Campaign. In: 83rd ISOLDE Collaboration meeting, Nov 2018. CERN-EDMS: 2363591/1
6. Fernandes S (2010) Submicro and nano structured porous materials for the production of high-intensity exotic radioactive ion beams. CERN-THESIS-2010-170. PhD thesis. CERN-CDS: 1312950
7. Fedosseev V, Chrysalidis K, Goodacre TD, Marsh B, Rothe S, Seiffert C, Wendt K (2017) Ion beam production and study of radioactive isotopes with the laser ion source at ISOLDE. J Phys G: Nucl Part Phys 44(8):084006
8. Kugler E, Fiander D, Johnson B, Haas H, Przewloka A, Ravn HL, Simon DJ, Zimmer K (1992) The new CERN-ISOLDE on-line mass-separator facility at the PS-Booster. Nucl Instrum Methods Phys Res Sect B 70(1):41–49
9. Kadi Y, Fraser MA, Papageorgiou-Koufidou A (2018) HIE-ISOLDE: technical design report for the energy upgrade. In: Kadi Y (ed) CERN yellow reports: monographs
10. Kugler E (2000) The ISOLDE facility. Hyperfine Interact 129(1–4):23–42
11. Habs D, Kester O, Bollen G, Liljeby L, Rensfelt KG, Schwalm D, Von Hahn R, Walter G, Van Duppen P (1997) The REX-ISOLDE project. Radioactive nuclear beams. Nucl Phys A 616(1):29–38
12. Habs D, Kester O, Rudolph K, Thirolf P, Hinderer G, Nolte E, Bollen G, Raimbault-Hartmann H, Ravn H, Ames F et al (1997) The REX-ISOLDE project. International conference on electromagnetic isotope separators and techniques related to their applications. Nucl Instrum Methods Phys Res Sect B 126(1):218–223
13. Wenander F, Jonson B, Liljeby L, Nyman GH (1998) REXEBIS the electron beam ion source for the REX-ISOLDE project. Tech. rep. CERN-OPEN-2000-320. CERN, Geneva. CERN-CDS: 478399
14. Borge MJG (2016) Highlights of the ISOLDE facility and the HIE-ISOLDE project. EPJ web of conferences, vol 117. EDP Sciences, p 10002
15. Forstner O (2001) Beam-preparation with REXTRAP for the REX-ISOLDE experiment. PhD thesis. Tech. U., Vienna. CERN-CDS: 520868
16. Ames F, Bollen G, Delahaye P, Forstner O, Huber G, Kester O, Reisinger K, Schmidt P (2005) Cooling of radioactive ions with the Penning trap REXTRAP. Nucl Instrum Methods Phys Res Sect A 538(1):17–32
17. Habs D, Kester O, Sieber T, Bongers H, Emhofer S, Reiter P, Thirolf PG, Bollen G, Aystö J, Forstner O et al (2000) The REX-ISOLDE project. Hyperfine Interact 129(1–4):43–66
18. Wenander F (2010) Charge breeding of radioactive ions with EBIS and EBIT. JINST 5(10):C10004
19. Wenander F (2002) EBIS as charge breeder for radioactive ion beam accelerators. Nucl Phys A 701(1–4):528–536
20. Ames F, Cederkäll J, Sieber T, Wenander FJC (2005) The REX-ISOLDE facility: design and commissioning report. CERN yellow reports: monographs. CERN, Geneva
21. Podlech H, Grieser M, Von Hahn R, Papureanu S, Repnow R, Schwalm D (1998) The 7-gap-resonator-accelerator for the REX-ISOLDE-experiment at CERN. Nucl Instrum Methods Phys Res Sect B 139(1–4):447–450

22. Von Hahn R, Grieser M, Habs D, Jaeschke E, Kleffner CM, Liebmann J, Papureanu S, Repnow R, Schwalm D, Stampfer M (1993) Development of seven-gap resonators for the Heidelberg high current injector. Nucl Instrum Methods Phys Res Sect A 328(1–2):270–274
23. Borge MJG, Riisager K (2016) HIE-ISOLDE, the project and the physics opportunities. Eur Phys J A 52(11):334
24. CERN. About HIE-ISOLDE. https://hie-isolde-project.web.cern.ch/about-hie-isolde.html. Accessed 11 Aug 2020
25. Williams L, Tock JP, Vandoni G, Delruelle N, Gayde JC, Bouzoud A, Leclercq Y, Pasini M (2011) Design of the High Beta Cryomodule for the HIE-ISOLDE Upgrade at CERN. CERN-ATS-2011-247. CERN-CDS: 1407543, 4 p
26. Kadi Y, Parchet D, Catherall R, Venturini Delsolaro W, Siesling E, Fraser MA, Blumenfeld Y, Calatroni S, Goddard B, Bernardes AP, Voulot D (2012) Status and future perspectives of the HIE-Isolde Project at CERN. Tech. rep. CERN-ATS-2012-251, MOOBA02. CERN-CDS: 1481516, 4 p
27. Gaffney LP. Private communication (unpublished)
28. Kinsey RR, Dunford CL, Tuli JK, Burrows TW (1996) The NUDAT/PCNUDAT Program for Nuclear Data. Paper submitted to the 9th International Symposium of Capture Gamma-Ray Spectroscopy and Related Topics, Budapest, Hungary, Oct 1996. Data extracted from the NUDAT database, version 2.8. https://www.nndc.bnl.gov/nudat2/. Accessed 4 May 2020
29. Catford WN (2002) Nucleon transfer studies with radioactive beams. Nucl Phys A 701(1–4):1–6
30. Butler PA, Cederkall J, Reiter P (2017) Nuclear-structure studies of exotic nuclei with MINI-BALL. J Phys G: Nucl Part Phys 44(4):044012
31. Wuosmaa AH, Schiffer JP, Back BB, Lister CJ, Rehm KE (2007) A solenoidal spectrometer for reactions in inverse kinematics. Nucl Instrum Methods Phys Res Sect A 580(3):1290–1300
32. Bildstein V, Gernhäuser R, Kröll T, Krücken R, Wimmer K, Van Duppen P, Huyse M, Patronis N, Raabe R, Collaboration T-REX et al (2012) T-REX. Eur Phys J A 48(6):85
33. Agostinelli S, Allison J, Amako KA, Apostolakis J, Araujo H, Arce P, Asai M, Axen D, Banerjee S, Barrand GJ, Behner F et al (2003) Geant4—a simulation toolkit. Nucl Instrum Methods Phys Res Sect A 506(3):250–303
34. Labiche M. Private communication (unpublished)
35. Lighthall JC, Back BB, Baker SI, Freeman SJ, Lee HY, Kay BP, Marley ST, Rehm KE, Rohrer JE, Schiffer JP et al (2010) Commissioning of the HELIOS spectrometer. Nucl Instrum Methods Phys Res Sect A 622(1):97–106
36. Lighthall JC (2011) Commissioning of the helical orbit spectrometer: a new device for measuring nuclear reactions in inverse kinematics. PhD Thesis, Western Michigan University. https://scholarworks.wmich.edu/dissertations/432. Accessed 10 May 2019
37. Micron Semiconductor Ltd (2020) S1. http://www.micronsemiconductor co.uk/product/s1/. Accessed 28 May 2020
38. ORTEC (2020) Si charged particle radiation detectors for research applications. https://www.ortec-online.com/products/radiation-detectors/silicon-charged-particle-radiation-detectors/si-charged-particle-radiationdetectors-for-research-applications. Accessed 1 Sept 2020
39. Micron Semiconductor Ltd (2020) QQQ1. http://www.micronsemiconductor.co.uk/product/qqq1/. Accessed 25 Aug 2020
40. Kamal A (2014) Nuclear physics, 1st edn. Springer, Berlin, Heidelberg

Chapter 5
Analysis of the $d(^{28}Mg,p)^{29}Mg$ Experiment

This chapter details the stages of analysis of the data obtained from the measurement of the $d(^{28}Mg,p)^{29}Mg$ reaction using ISS. The procedures required to extract absolute cross sections and spectroscopic factors for the states populated in the $d(^{28}Mg,p)^{29}Mg$ reaction are described. This includes details on calibration of the raw data and selection of the reaction of interest, as well as methods for making ℓ-assignments for each state.

5.1 Detector Calibrations

5.1.1 Gain Matching

As detailed in Sect. 4.6.1, the HELIOS silicon array records three signals: X_1, X_2, and E. To account for variations in signal gain and offsets between the 24 detectors in the array, all of these signals must be gain matched for each of the 24 detectors on the silicon array.

The detectors were gain matched using α particles of known energies from a mixed α-source, containing the four radionuclides listed in Table 5.1. The α particles were emitted from rest in the lab frame, so every α particle from a given decay deposited the same energy on the array, independent of position. One of the raw α-particle energy spectra is shown in Fig. 5.1.

These four strongest α-particle peaks can be clearly distinguished, therefore providing four data points from which a robust calibration can be created. A linear relationship of the form

$$E = a(X_1 + bX_2) + c \tag{5.1}$$

was used to gain match the three signals for each detector. a, b, and c are gain-matching constants particular to each detector on the array. Gain matching was achieved in a number of stages. First, the signals X_1 and X_2 were matched to each

P. T. MacGregor, *Single-Particle Structure of ^{29}Mg on the Approach to the N=20 Island of Inversion*, Springer Theses, https://doi.org/10.1007/978-3-031-19119-0_5

Table 5.1 The four radionuclides used in the mixed α-source, and the energies of the strongest decay branch [1] for each

Radionuclide	Energy (keV)	Radionuclide	Energy (keV)
^{148}Gd	3183	^{239}Pu	5157
^{241}Am	5486	^{244}Cm	5805

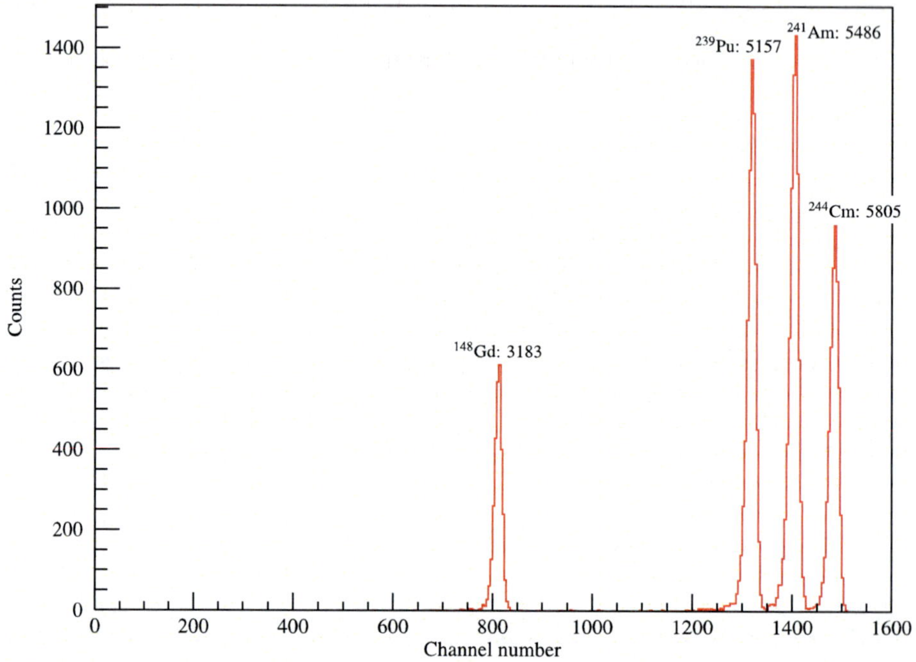

Fig. 5.1 An α-particle spectrum from one of the detectors on the array. Each peak has been labelled with the α-particle energy in keV

other by scaling X_2 by b. Then, this weighted sum of X_1 and X_2 was scaled by a to match the raw energy signal, E. Any offset required was contained in c.

Figure 5.2 shows the process of extracting these gain-matching parameters. Alpha particles from the mixed source produced four distinct α-particle lines on the raw X_1–X_2 plot, which is shown in Fig. 5.2a. Each line has the same gradient, which allowed the extraction of the parameter b from one of the lines, shown in black. The parameters a and c were then extracted from the fit between E and $X_1 + bX_2$, shown in Fig. 5.2b.

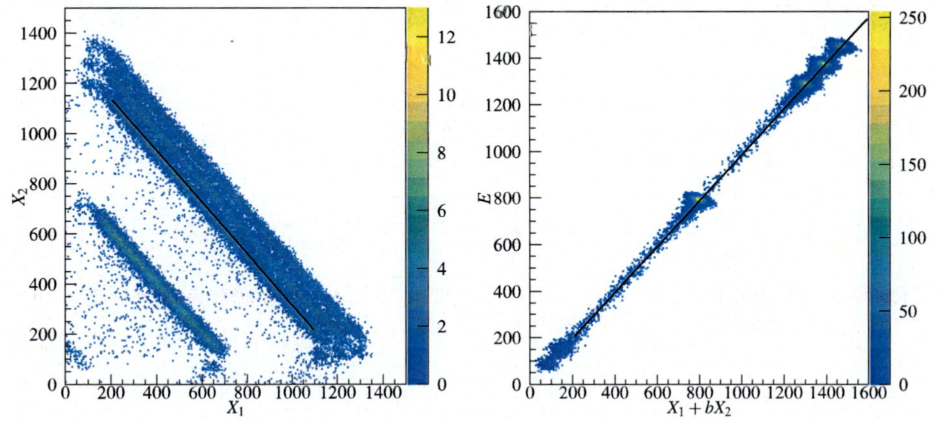

(a) (b)

Fig. 5.2 **a** The plot of the raw X_1 and X_2 signals. The four α-lines can be distinguished. **b** The matched X_1 and X_2 compared to the raw E signal. The black lines indicate the fits used to extract the parameters a, b, and c

5.1.2 *Energy Calibration*

A further linear calibration of the deduced excitation energy spectrum was performed using energies from Ref. [2] for each detector. This accounted for potential offsets in the calibration due to energy losses of the emitted α particles in the source.

5.1.3 *Contributions to Excitation-Energy Resolution on the Array*

The observed excitation-energy resolution depends on a number of factors.

- Proton energy resolution: this was discussed in Sect. 4.6.1, and is 27–53 keV at FWHM, depending on the detector [3].
- Energy losses of beam or ejectile in the target will contribute to the energy spread of protons depending on where the reaction occurs in the target.
- *Energy loss of the beam particle*: the incoming ^{28}Mg is slowed by interacting with the target material. This was calculated using stopping tables [4]. For a nominal target thickness of 100 μg cm^{-2}, the maximum energy loss to the beam is 610 keV if it fully passes through the target. This energy loss scaling is approximately linear with the depth of the reaction site within the target. This corresponds to a 13 keV maximum change in the ejectile energy for $\eta_{lab} = 0°$ for protons.

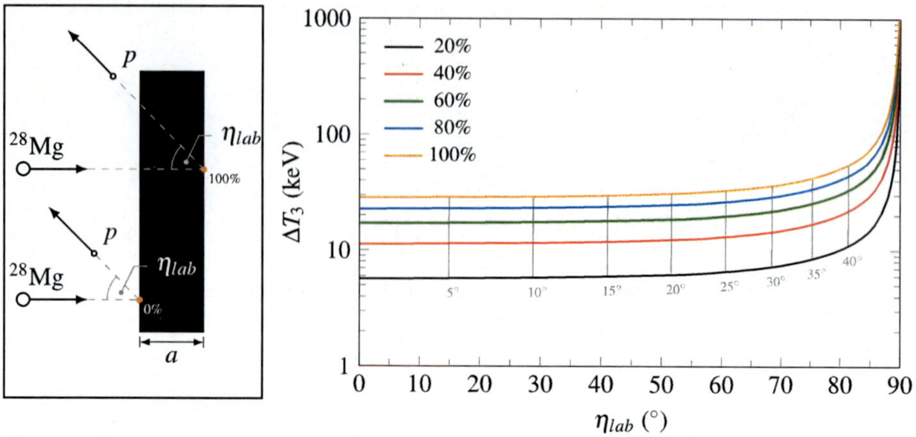

Fig. 5.3 Left: illustrations of the penetration depth within the target. A greater percentage of penetration causes greater energy loss to the ejectile. Right: the maximum change in ejectile energy at a given depth and emission angle for the ground state compared to an ejectile that reacts on the surface of the target and experiences no energy loss, ΔT_3. This is shown for different depths of penetration in a 100 μg cm^{-2} CD_2 target quoted as a percentage of the full target thickness. Corresponding η_{cm} values have been labelled for each penetration

- *Energy loss of the ejectile*: the proton experiences energy loss passing through the target. This is dependent on the depth of the reaction site, but also on the angle of emission. Protons emitted at high values of η_{lab} lose more energy to the target, and this energy loss is proportional to $\frac{a}{\cos \eta_{lab}}$, where a is the target thickness. The maximum energy loss is 15 keV for a proton passing through a 100 μg cm^{-2} target where $\eta_{lab} = 0°$.
- These effects are illustrated in Fig. 5.3, showing the change in ejectile energy for a 100 μg cm^{-2} target. For the targets used here, with nominal thicknesses of 80 and 120 μg cm^{-2}, the maximum energy lost by the ejectile protons was 23 keV and 34 keV, respectively, at $\eta_{lab} = 0°$.

- Beam energy spread: the beam energy was 9.473 MeV/u on average, but had an energy spread of $\frac{\delta E}{E} = 0.3\%$ [5]. This corresponds to a beam energy spread of 28 keV/u FWHM, and leads to a spread in ejectile energy of 5–14 keV for $0° \leqslant \eta_{cm} \leqslant 30°$ for protons populating the ground state in $d(^{28}Mg,p)$.
- Detector alignment: each detector within ISS was precisely aligned within the solenoid with uncertainty less than 0.1 mm [6]. Therefore, the alignment was taken to have negligible effects on the resolution.

Similarly, there are effects due to the position resolution of the detectors. Firstly, the position resolution affects the energy resolution. As mentioned in Sect. 4.6.1, the energy dependence on z contributes 16.4 keV/mm to the energy resolution. For the array, z is determined within 1 mm, so contributes 16.4 keV at most to the resolution [3].

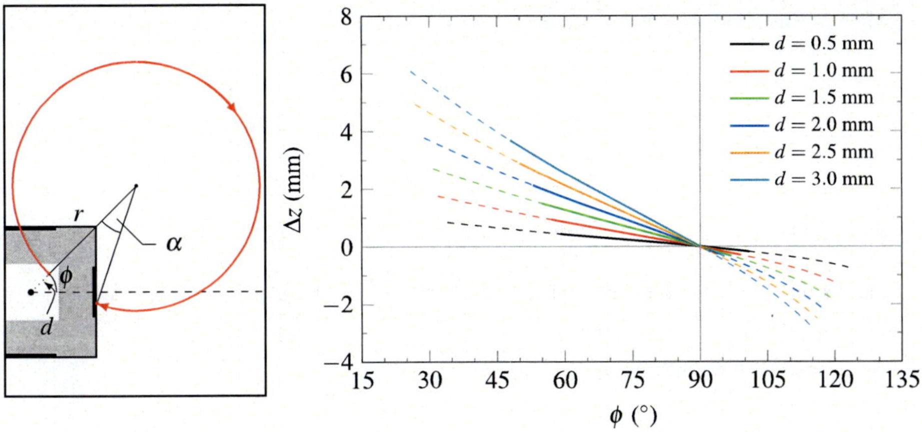

Fig. 5.4 Left: the orbit of an ejectile that is displaced a distance d from the centre of the array. This orbit is characterised by its radius, r, azimuthal emission angle, ϕ, and by the angle between the start and end position measured from the centre of its orbit, α. Right: the effect of changing the distance d and the angle ϕ on z. The term Δz is the change in z from the case where $d = 0$ mm for a proton that is ejected from a reaction where $\eta_{cm} = 11°$, and populates the ground state. This is limited to one side of the array. The solid lines show protons that hit a detector, and the dashed lines show the protons that hit the side of the array but miss the detector

However, this is further compounded by the size of the beam spot, the effects of which are shown in Fig. 5.4. A finite beam spot causes ejectiles to be emitted from the target off-centre. This changes where they will hit the array detectors after completing their helical orbit, leading to a range of values for z for a given T_3. This can be characterised with the angle, α (see Eq. 4.22), as shown in the left panel of Fig. 5.4.

The broadening of the possible values of z for a given excitation energy leads to increased FWHM of the peaks in the excitation energy spectrum. This is minimised at $\phi = 90°$ in this particular case, as α remains constant irrespective of the displacement from the centre of the detector. Higher excitation energies and higher values of η_{cm} have reduced broadening, as the range of possible z values is reduced due to a smaller value of $u_{\parallel}$. What is shown in Fig. 5.4 is therefore the maximum possible broadening in z for the range of angles considered.

For this experiment the beam spot size was less than 2 mm FWHM, having been tuned through a 3 mm collimator. For $d = 1.0$ mm, most of the ejectiles have a change in z within the position resolution of the detector. Nevertheless, this contributes approximately 33 keV to the energy resolution.

Combining these resolutions gives an expected, excitation-energy resolution of 100–160 keV for each detector. The measured FWHM of the first three states is shown in Table 5.2 for each row of detectors, where the four detectors with the same z have been summed together. The ground state doublet was present in all detectors, and is given as a comparison to the isolated states at 1092 and 1492 keV where possible.

Table 5.2 The width of the two strong excited states at 1092 and 1492 keV, and the width of the ground-state doublet for comparison

Row	FWHM (keV)	
	1092, 1432	0 + 55 doublet
1	–	164(14)
2	182(8)	176(5)
3	144(3)	159(5)
4	139(3)	182(6)
5	154(4)	158(8)
6	158(6)	174(14)

This was measured for the 80 μg cm^{-2} target

5.2 Calculating the Target and Beam Properties for Extracting Absolute Cross Sections

The quantity $I_0 N_t$, the product of the beam intensity and the number of target particles per unit area, is necessary for extracting the cross sections for populating states in ^{29}Mg, and was determined using deuteron elastic scattering. Rearranging Eq. 3.2 gives

$$I_0 N_t = \frac{Y}{\frac{d\sigma}{d\Omega}\Delta\Omega\varepsilon_\pi\varepsilon_\delta}. \tag{5.2}$$

The yield, Y, corresponds to the number of elastically-scattered deuterons that were incident on the luminosity detector.

The efficiency of the detector, ε_δ, was assumed to be 1, because the combination of a relatively low beam current and a small solid angle of acceptance ensures that there is little dead time.

Elastically-scattered deuterons from reactions with ^{28}Si are incident at effectively identical angles and energies to those from reactions with ^{28}Mg. The purity of the beam, ε_π, was determined using the ΔE–E beam diagnostic detector. Data from this detector are shown in Fig. 5.5, where the ^{28}Si contaminant can be clearly seen with the ^{28}Mg beam. Comparing the yield of ^{28}Mg to ^{28}Si yields a beam purity of 0.925 ± 0.007.

Since the solid angle was calculable (as described in Sect. 4.6.2), and the yield was measured, the remaining quantity needed to determine $I_0 N_t$ was the cross section. At the beam energy used, the scattered deuterons cannot be described using Rutherford scattering, so the cross section was calculated using the optical model for a range of different global optical-model parameters listed in Table 5.3. The ratio-to-Rutherford cross sections for these different parameter sets are presented in Fig. 5.6. The mean ratio was 81.2% with a relative difference between the models of $\pm 0.7\%$.

As the luminosity detector has a finite angular acceptance, an integrated cross section was used:

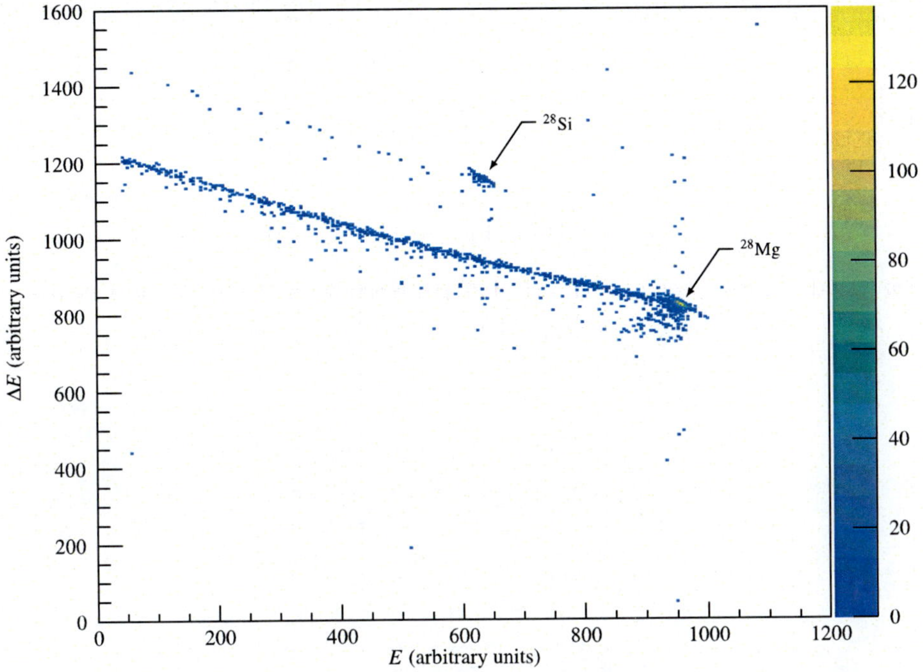

Fig. 5.5 The beam purity indicated by the ΔE–E detector. Two distinct particle groups are present showing the desired ^{28}Mg beam and the contaminant ^{28}Si. The "tails" on each particle group arise due to scattering of the beam

$$\left\langle \frac{\mathrm{d}\sigma}{\mathrm{d}\Omega} \right\rangle = \frac{\int \frac{\mathrm{d}\sigma}{\mathrm{d}\Omega} \mathrm{d}\Omega}{\int \mathrm{d}\Omega} = \frac{\int \frac{\mathrm{d}\sigma}{\mathrm{d}\Omega} \mathrm{d}\Omega}{\Delta\Omega}, \tag{5.3}$$

so Eq. 5.2 becomes

$$I_0 N_t = \frac{Y}{\int \frac{\mathrm{d}\sigma}{\mathrm{d}\Omega} \mathrm{d}\Omega}. \tag{5.4}$$

This integral was calculated numerically. However, the luminosity detector had significantly lower yields of deuterons for two of its four quadrants. This was attributed to the potential blocking of trajectories between the target area and the detector due to the finite thickness of the target ladder. Therefore, only two quadrants were used to extract the yield. An example of the measured yield is shown in Fig. 5.7.

The peak here lies on the background of lower energy events. They primarily occur due to scattering from the ^{12}C and noise on the detector, but there are contributions from deuterons that undergo multiple turns and particles from fusion evaporation reactions. This background was estimated with a quadratic fit.

Two targets were used for the experiment, characterised in Table 5.4. Though Position 2 of the array had a thicker target, and hence a larger value of N_t, the overall

Table 5.3 Details of the optical model parameters used for calculating the $d(^{28}Mg,d)^{28}Mg$ reaction cross section

	Real volume			Imaginary volume			Imaginary surface			Real spin orbit			Imaginary spin orbit			R_C	References
	U_v	$r_{0,v}$	a_v	W_v	$\tilde{r}_{0,v}$	$\tilde{a}_v$	W_s	$\tilde{r}_{0,s}$	$\tilde{a}_s$	U_{so}	$r_{0,so}$	a_{so}	W_{so}	$\tilde{r}_{0,so}$	$\tilde{a}_{so}$		
(a)	89.711	1.149	0.757	2.282	1.338	0.551	10.25	1.384	0.719	3.557	0.972	1.011	0	0	0	1.303	[8]
(b)	82.434	1.18	0.742	0	1.27	0.832	10.958	1.27	0.832	6	0.895	0.895	0	0	0	1.3	[9]
(c)	87.052	1.17	0.741	0.448	1.325	0.742	12.245	1.325	0.742	6.781	1.07	0.66	0	0	0	1.3	[10]
(d)	86.116	1.17	0.74	0.444	1.332	0.732	12.144	1.332	0.732	6.594	1.07	0.66	0	0	0	1.3	[10]
(e)	78.266	1.174	0.809	1.151	1.563	0.837	13.264	1.328	0.602	3.703	1.234	0.813	−0.206	1.234	0.813	1.698	[11]
(f)	99.824	1.05	0.86	0	0	0	23.642	1.43	0.62	7	0.75	0.5	0	0	0	1.3	[12]
(g)	84.735	1.15	0.81	0	0	0	18.947	1.34	0.68	0	0	0	0	0	0	1.15	[13]

The symbols are defined in Sect. 3.5, with symbols with ~ denoting quantities accompanying imaginary potentials. The unit for the potentials is MeV, and the unit for the radius parameter and the diffuseness is fm

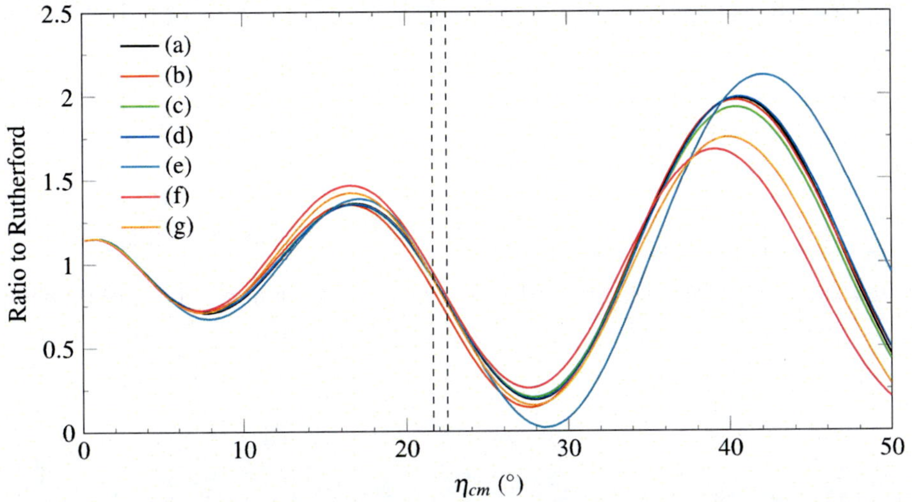

Fig. 5.6 The ratio of the DWBA cross section to the Rutherford value for the $d(^{28}Mg,d)^{28}Mg$ reaction at 9.473 MeV/u. The different optical models used are listed in Table 5.3. The dashed lines correspond to the angular acceptance of the luminosity detector for a single orbit. The optical-model cross sections were calculated using PTOLEMY [7]

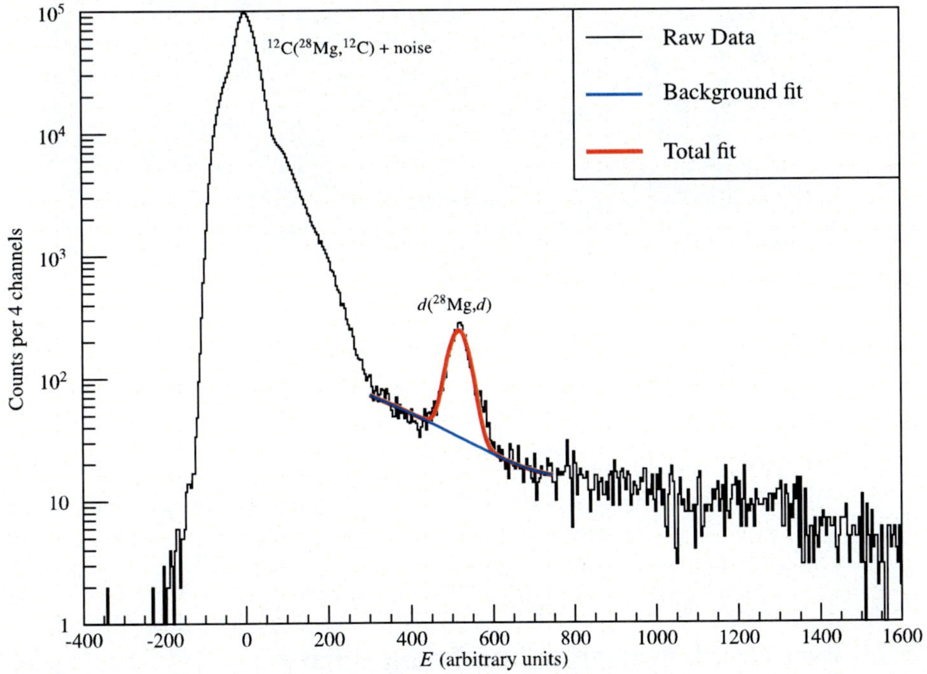

Fig. 5.7 A plot of the extracted yield from the luminosity detector. A Gaussian peak was fitted on a quadratic background (blue) to form a fit for the desired deuterons (red)

Table 5.4 Details of the targets used for extracting cross sections

Array position	Nominal thickness ($\mu g\ cm^{-2}$)	$I_0 N_t \pm$ stat.	sys. $\pm$ mb^{-1}
1	80	743 ±15	± 76
2	120	365 ± 11	± 41

The statistical error in $I_0 N_t$ arises from the error on extracting the yield, whereas the systematic error arises from both the optical model dependence and uncertainty in the array position, ±5 mm. Uncertainties are discussed further in Sect. 5.7

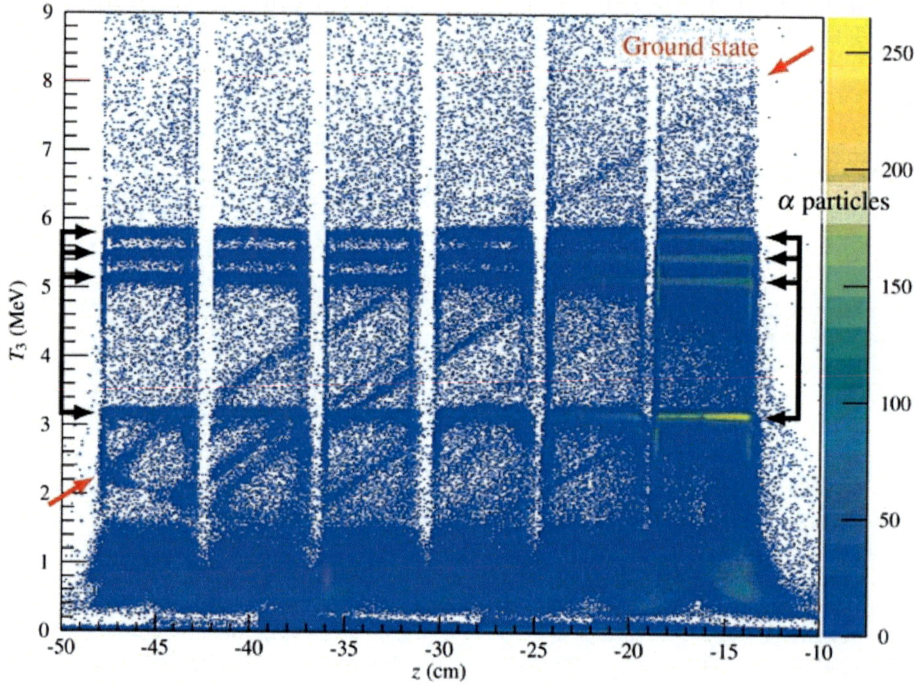

Fig. 5.8 Measured ejectile energy, T_3, plotted against z along the detector array. No software gates have been applied to the data

value of $I_0 N_t$ was less because the measurements were taken at this position for a shorter period of time. The dependencies of the extracted absolute cross sections and the value of $I_0 N_t$ on the choice of optical model is discussed further in Sect. 5.9.

5.3 Particle Identification and Beam Purity

The raw data, before any software gates are applied, are shown in Fig. 5.8. This histogram shows all of the detected particles on the array. The four α-particle lines can be seen horizontally here, as the quadruple α-source remains within the solenoid

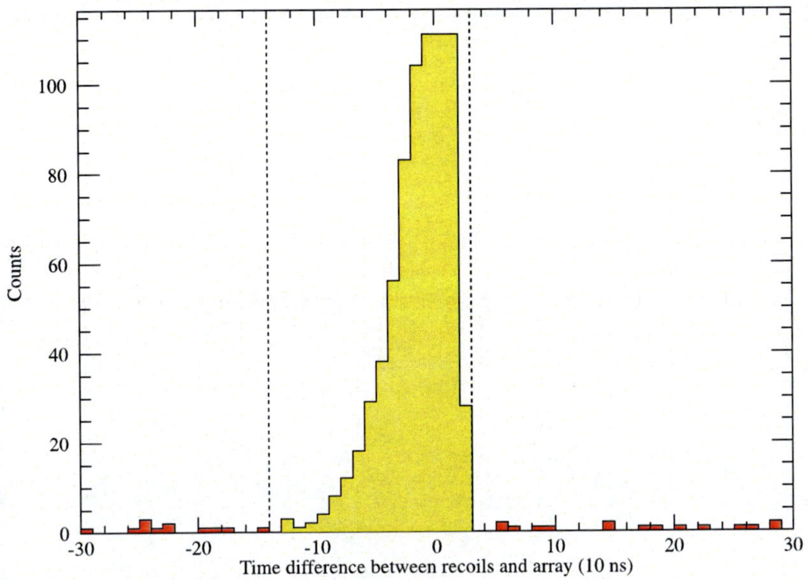

Fig. 5.9 The time difference cut for a detector on the array. The accepted data are between the dashed lines (yellow). Outside of this region, the data are rejected (red)

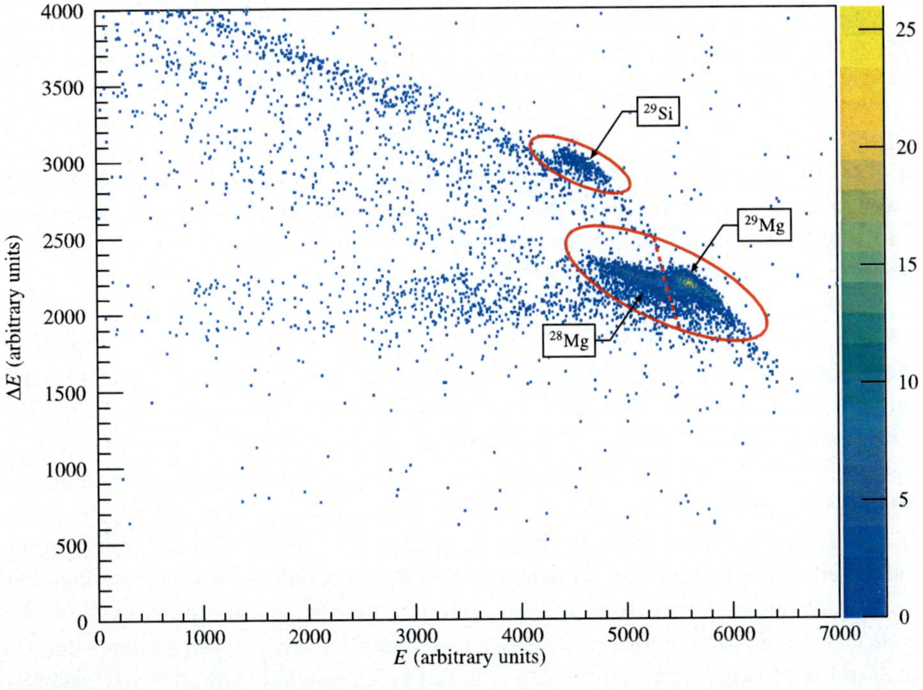

Fig. 5.10 An example of the data collected in one quadrant of the recoil detector with time gates applied to select for coincidences. The main concentration of Mg nuclei and Si nuclei have been highlighted. Each nuclide has a low-energy "tail" corresponding to elastic scattering. The dashed red line indicates the approximate division between the ^{29}Mg and ^{28}Mg particle groups

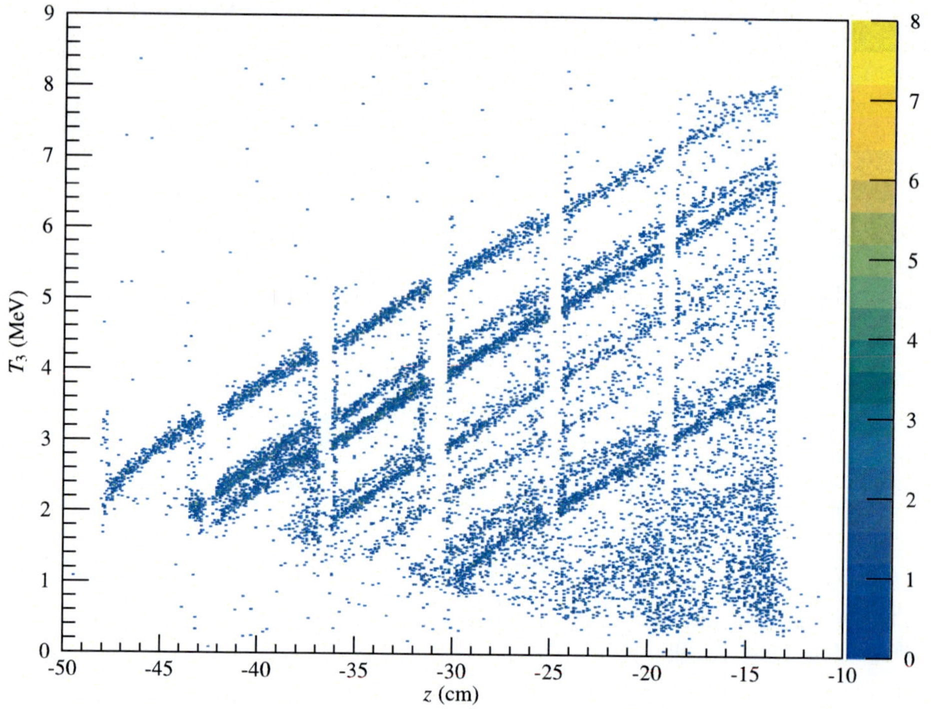

Fig. 5.11 Measured ejectile energy T_3 against z. Timing cuts and gating on the recoil detector have been implemented to select for magnesium nuclei. The gaps between silicon detectors on the array are visible on this plot

and all of its emissions are not blocked. The protons, present where a low-lying excited state in ^{29}Mg has been populated, can be seen as diagonal lines resembling those in Fig. 4.15. They are significantly obscured by other artefacts, so software gates are necessary here to remove the unwanted data.

When a particle hits a detector in the ISS, the resulting signal triggers the digitisers on the data acquisition, where both the signal and a timestamp are recorded. In Sect. 4.4.1, it was shown that particles hitting the array could be identified by their time period to complete one orbit. Whilst this cannot be measured directly, the time difference between detection of a recoil and a proton provides identification of the reaction itself. This is possible because the variation in recoil timings is small; the maximum variation, arising from different values of η_{cm}, is less than 1.7 ns.

Therefore, the time difference between particles on the recoil detector and protons on the array can be constrained with a software time window, and events that lie outside of this window were rejected. An example time spectrum is shown in Fig. 5.9. A gate around the peak in this spectrum imposes a particle-recoil coincidence condition within the selected time difference. This can be calculated using the time taken for the recoiling ^{29}Mg to reach the recoil detector, and the cyclotron period of the proton, and is approximately 8 ns.

This reduces background events, including protons from fusion evaporation reactions that fall outside the timing window, and α particles from the calibration source.

However, not all protons within this time window arise from a reaction with ^{28}Mg. Nonetheless, due to the presence of beam contaminants (primarily ^{28}Si), recoil nuclei can be distinguished because they have different energy loss characteristics in the ΔE–E recoil detectors. Applying software cuts to the signals from the ΔE and E detectors removes reactions with contaminants in the beam. An example of the recoil-detector spectrum is shown in Fig. 5.10.

The two-dimensional cuts were optimised by placing gates around the desired particle group, and slowly reducing the gate to eliminate background while still maintaining the yield in the peaks of the excitation spectrum. The effects of these cuts on the T_3–z plot is shown in Fig. 5.11.

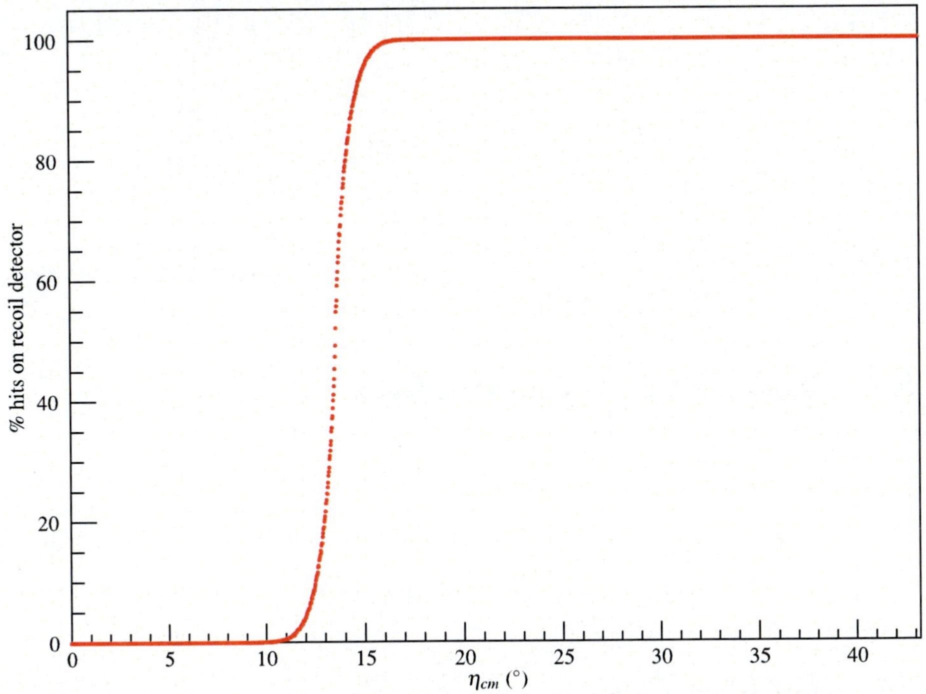

Fig. 5.12 A simulation showing the percentage of residual nuclei that hit the recoil detector given a detection of the ejectile. The angular range covers all ejectiles that complete up to one helical orbit when $E_x = 0$ MeV. 20,000 events were used to determine the percentage for each data point. The value of ϕ in each event was pseudo-random

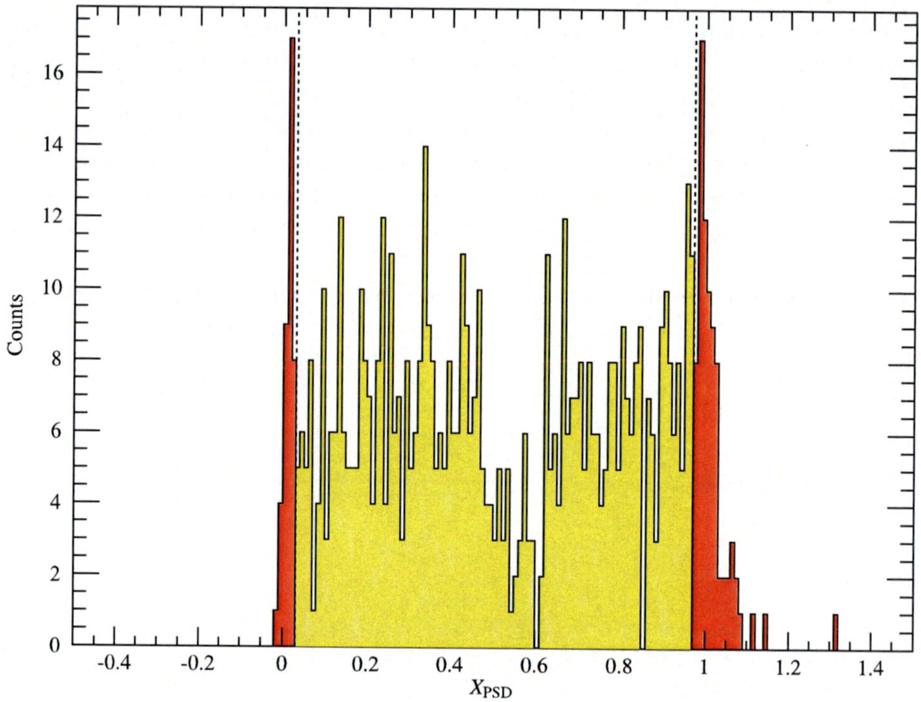

Fig. 5.13 The cuts (red) on a given detector on the array. The term X_{PSD} corresponds to the value of z on a PSD. It is normalised to be between 0 and 1

5.4 Additional Cuts to Clean the Data

A series of additional cuts were implemented to further improve the signal-to-background in the excitation spectrum, and they are detailed in the following subsections. The effect that each of these cuts in turn had on the data is detailed in Appendix C.

5.4.1 CM Angle Cuts

Cuts in η_{cm} are required to remove the "knees" from the T_3-z plot, where the kinematic lines become double-valued due to the finite size of the detector array. This occurs at $\eta_{cm} \approx 11°$. However, an implicit angle cut is already imposed by the selection of ^{29}Mg in the recoil detector. Here, ejectiles that hit the array are forced to have a corresponding recoil nucleus that also hits the recoil detector. This is independent of the position of the "knees".

The position of these limits can be simulated, and an example for the ground state is shown in Fig. 5.12. This shows the percentage of recoiling ^{29}Mg that hit the recoil

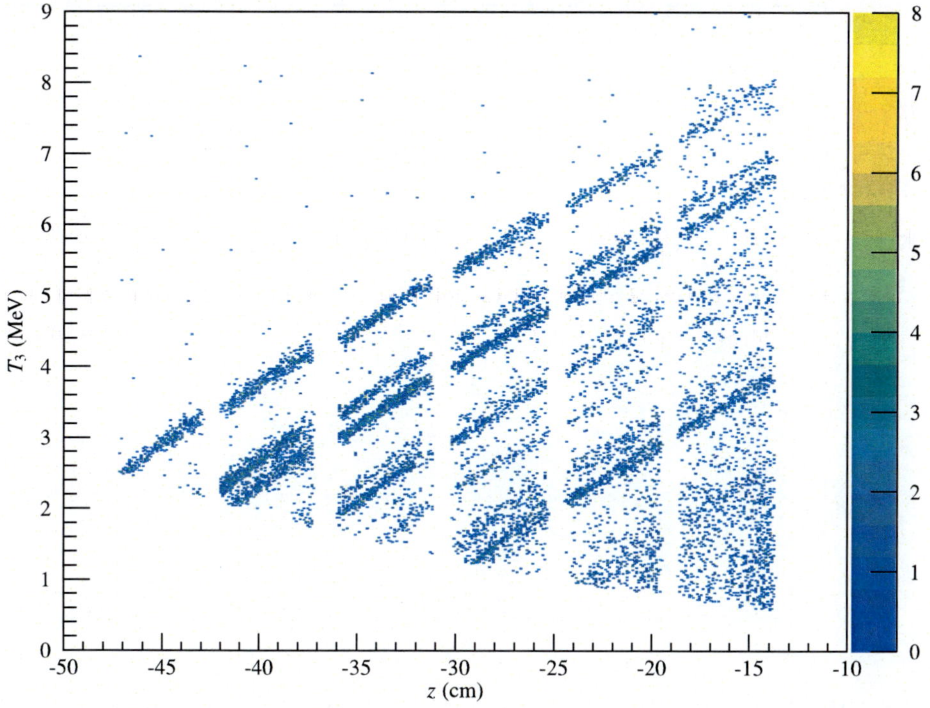

Fig. 5.14 The final T_3-z plot for the $d(^{28}\mathrm{Mg},p)^{29}\mathrm{Mg}$ reaction. The cuts used are described in the main text

detector for a given η_{cm}. A threshold can be implemented on a state-by-state basis where 99% of the recoils hit the detector. This ensures that the calculated solid angles are not overestimated when extracting the yields. However, these cuts remove the forward-angle data from the final excitation spectrum, so these data were considered separately (see Sect. 5.5).

5.4.2 Applying Cuts on the Detector Position

Several PSDs showed unphysical peaks in the position spectra at their extremes, as shown in Fig. 5.13 in red. These could arise from noise on either position signal, X_1 or X_2, which artificially skews the calculated position to favour values closer to 0 or 1. Additionally, there could be effects arising from the guard ring that surrounds the detector. It was in electrical contact with the bonding pads, and is designed to control the shape of the electric field in the detector [3]. However, any particles that land on the guard ring may induce a signal on the bonding pad which is used for one of the position signals.

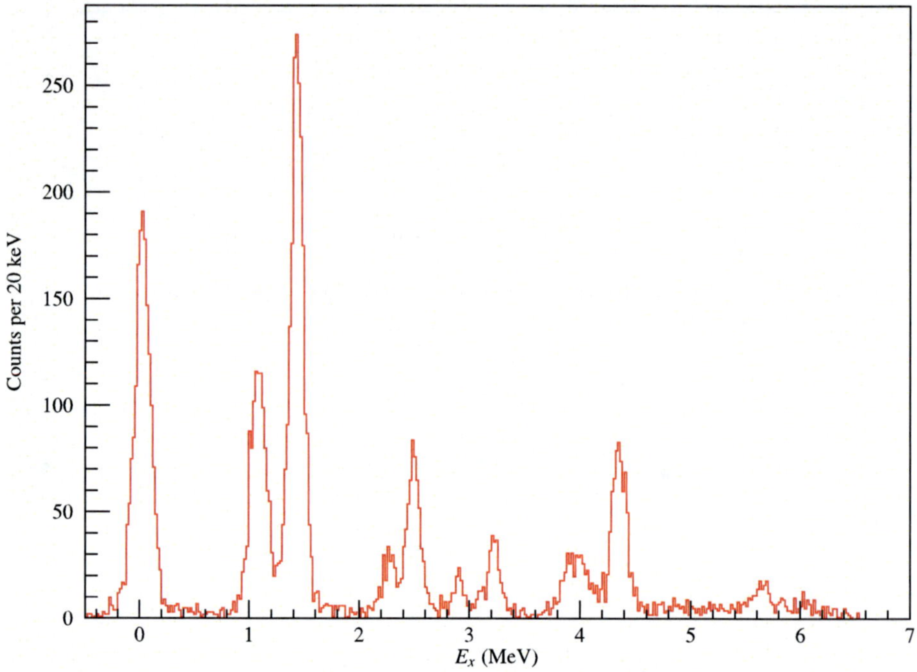

Fig. 5.15 The final ^{29}Mg excitation-energy spectrum for protons in coincidence with recoils. The cuts used are described in the main text

Therefore, to counteract this effect, cuts were placed to remove these events, and X_{PSD} was limited to $0 \leqslant X_{PSD} \leqslant 1$ at the very most. This was the final cut placed on the data.

After these cuts had been applied, the T_3-z plot (see Fig. 5.14) closely resembled the schematic plot in Fig. 4.15. The corresponding excitation-energy spectrum is shown in Fig. 5.15.

5.5 Extracting Cross Sections of Forward Angles Using Singles Data

As mentioned in Sect. 5.4.1, data at forward angles were excluded for events involving coincidences with recoils. Therefore, to obtain cross sections at these angles, singles events from the silicon array were used.

However, extracting cross sections using these data was complicated due to the presence of the fusion-evaporation background, competing transfer reactions, and α particles. These issues have been highlighted in Fig. 5.16 showing the singles data with no cuts applied.

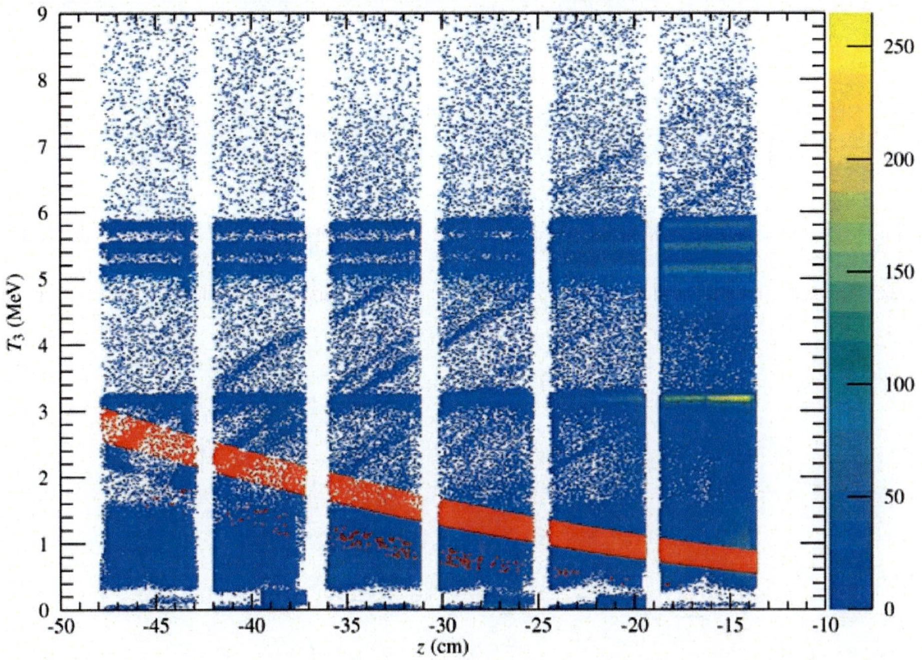

Fig. 5.16 The singles data on a T_3-z plot where no cuts have been applied. The highlighted region in red is for data where $11.0 \leqslant \eta_{cm} \leqslant 14.5°$. Some events outside of this band are highlighted, showing where the calculation (see text for details) has failed to converge to the true minimum. However, any contribution from these events is insignificant

In order to avoid contamination from the α particles, narrow η_{cm} cuts were required to acquire select portions of the desired events.[1] This is illustrated for obtaining information on the ground state in the first row of detectors at low η_{cm}. The chosen cut was designed to remove the knees from the data, while also avoiding the first α-particle line.

However, the lack of recoil gating means contaminants are still present in the excitation spectrum. Therefore, software gates were drawn around the most prominent contaminant particle group in the recoil detector spectrum, corresponding to ^{28}Si (see Fig. 5.10). A total excitation spectrum was then drawn, and is compared to the corresponding spectrum for ^{29}Mg in Fig. 5.17.

Here, there are two states of ^{29}Si that significantly overlapped states in ^{29}Mg. These had excitation energies of 7997 keV, which had been previously identified, and 9357 keV, which had not. States at 6.6 and 11.4 MeV were also observed, but these do not significantly overlap a state of ^{29}Mg.

Due to low yields overall, it was only possible to extract cross sections integrated across all of the detectors on the array, σ_{Si}. Then, using a similar cross section for the

[1] Values of η_{cm} are calculable for each measurement of ejectile energy and z—see Appendix A for further details.

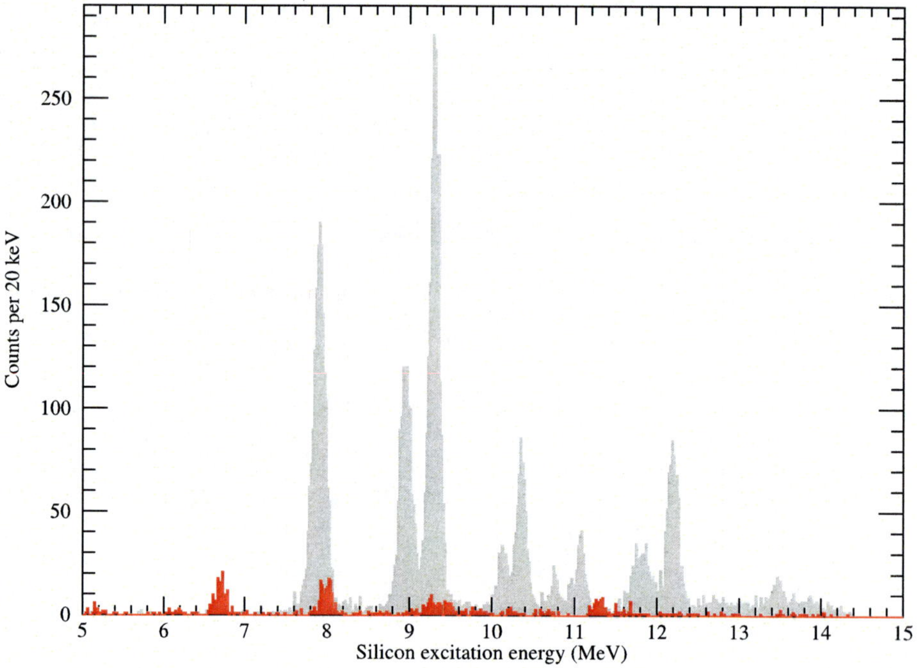

Fig. 5.17 The ^{29}Si excitation spectrum (red) with the ^{29}Mg excitation spectrum in terms of the ^{29}Si excitation energy (grey). The ^{29}Si spectrum was produced using a recoil-detector gate similar to the one drawn in Fig. 5.10. The ground state of ^{29}Mg occurs at just under 8 MeV in the ^{29}Si excitation energy

^{29}Mg (σ_{Mg}), the magnesium differential cross section at forward angles was estimated with

$$\left.\frac{d\sigma}{d\Omega}\right|_{\text{Mg}} = \frac{\sigma_{\text{Mg}}}{\sigma_{\text{Si}} + \sigma_{\text{Mg}}} \left.\frac{d\sigma}{d\Omega}\right|_{\text{sing.}} . \tag{5.5}$$

The fraction in this equation is the ratio of the contaminant cross section integrated across the detectors to the total of silicon and magnesium. This weights the individual cross sections obtained from the forward-angle data where silicon contaminants are significant, in order to estimate the differential cross section at these angles.

5.6 Calculating Solid Angles

Solid angles were calculated on a detector-by-detector basis for each individual state. As shown in Sect. 4.6.1, the solid angle for a single detector was shown to be

$$\Delta\Omega = \frac{\Delta\phi}{v_3 t_{cyc}} \Delta z = \Delta\phi \left(\cos\eta_1 - \cos\eta_2\right) . \tag{5.6}$$

The angles η_1 and η_2 are the polar angular limits of the detector, where an ejectile of the desired excitation energy must have an angle within this region in order to hit the detector. The corresponding limits in z are z_1 and z_2, where $\Delta z = z_2 - z_1$.

Due to the applied position cuts, the value of Δz varies detector-by-detector. Additionally, η_{cm} cuts impose a z condition on the same detectors. Using these two constraints, values of η_1 and η_2 were calculated on a detector-by-detector basis.

As the yields were extracted for a given row of detectors (groups of four with the same z), the total solid angle for that row was the sum of each of the individual solid angles for each detector.

The angle η_{cm} is non-linear in z, so it is not strictly accurate to take either the midpoint of z for the detector and convert that to an angle or the average of η_1 and η_2 as the centroid. Instead, the centroid of each detector, $\overline{\eta}$ can be calculated using an average of the total angle spanned by the detector that is weighted by the solid angle coverage. Mathematically, this is expressed as

$$\overline{\eta} = \frac{\int \eta \, \mathrm{d}\Omega}{\int \mathrm{d}\Omega} = \frac{\int_{\eta_1}^{\eta_2} \eta \sin \eta \, \mathrm{d}\eta \int_{\phi_1}^{\phi_2} \mathrm{d}\phi}{\int_{\eta_1}^{\eta_2} \sin \eta \, \mathrm{d}\eta \int_{\phi_1}^{\phi_2} \mathrm{d}\phi} = \frac{\sin \eta_2 - \sin \eta_1 + \eta_1 \cos \eta_1 - \eta_2 \cos \eta_2}{\cos \eta_1 - \cos \eta_2} \tag{5.7}$$

for a particular detector. This was found for a given row to give the corresponding centroid. In practise, while the difference is largest for small angles, the difference between this and the mean of the two angles is within uncertainties.

5.7 Extracting Cross Sections

The yields for each peak were extracted using the `gf3` peak-fitting software package [14]. A series of Gaussian functions were fitted to the peaks. For most of the peaks corresponding to bound states, the widths for weaker states were fixed based on the widths of strong singlets. For most of the peaks corresponding to unbound states, the widths were constrained to have the bound-state width as their lower limit. An excitation energy spectrum identifying the states is shown in Fig. 5.18.

A flat background was assumed for this spectrum as other background contributions have been removed by the cuts. The remaining background counts were assumed to come from imperfections in the gating.

Contributions to the overall statistical uncertainty of the cross sections include:

- *Uncertainty on the luminosity in* $I_0 N_t$—this was previously discussed in Sect. 5.2, and amounts to 3%.
- *Uncertainty in the detector solid angle*, $\Delta\Omega$—this arises from the position resolution of z. This uncertainty in z was ± 1 mm. Values of $\Delta\Omega$ were therefore calculated at these offset values of z, and the standard deviation of the three values

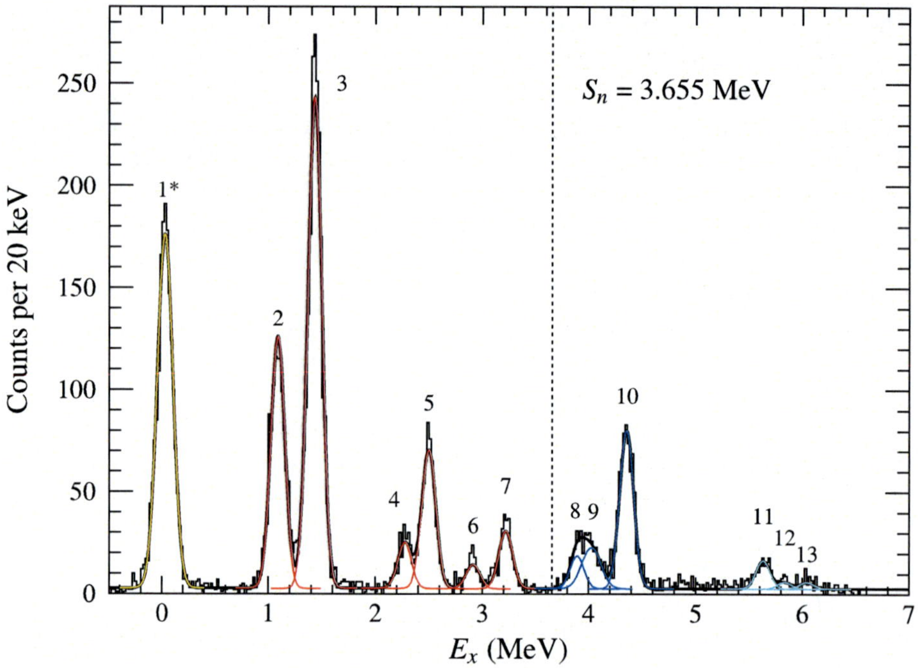

Fig. 5.18 The excitation spectrum, with Gaussian peaks highlighting the peaks where yields were extracted. The neutron separation energy, S_n, has been highlighted, above which states in ^{29}Mg are unbound

of $\Delta\Omega$ was taken to be the uncertainty in $\Delta\Omega$. However, this uncertainty is of the order of 0.01%, so can be neglected.

- *Uncertainty in the measured yield for each detector*—this is by far the largest contributor to the uncertainty in the cross section, and is typically of the order 5–15% for array position 1, and 5–30% for position 2.

All of the statistical uncertainties were appropriately combined to obtain a final uncertainty in the cross section.

Systematic uncertainties are dominated by those arising from the luminosity, $I_0 N_t$, amounting to 10%.

5.8 Assigning Angular Momenta and Extracting Spectroscopic Factors

5.8.1 Previous Assignments for ^{29}Mg

Before the current work, several excited states in ^{29}Mg have been identified and assigned angular momentum quantum numbers in a number of previous experimental studies. The most recent study by Matta et al. [15] summarises the known states of ^{29}Mg from a variety of experiments, and contributes data from a $d(^{28}\mathrm{Mg},p\gamma)^{29}\mathrm{Mg}$ performed at the TRIUMF facility with the SHARC silicon strip detector array [16] for detecting protons in coincidence with γ rays measured in 12 TIGRESS clover germanium detectors [17].

However, without the benefits of a solenoid, these measurements were affected by the effects of kinematic shift and compression, such that the total resolution of the proton energy was 700 keV FWHM. In addition, the beam was heavily contaminated: only 1% or 3000 pps were of the desired ^{28}Mg, which is a relatively small beam current. The proton resolution was mitigated due to the improved γ-ray energy resolution, which was 42 keV. However, the low statistics still limited this measurement. With previous assignments to help identify states, it was possible for Matta et al. to extract spectroscopic factors for a total of seven states, though some of these were extracted by fitting a combination of DWBA calculations of two different ℓ transfers to the experimental data as some states remain unresolved [15].

A number of other studies have examined the states in ^{29}Mg, using other techniques. These are:

- β decay of ^{29}Na [18].
- β-γ coincidences in ^{29}Na [19].
- β-$n\gamma$ decay of ^{30}Na [20].
- β-delayed spectroscopy of a polarised ^{29}Na beam [21].
- Single-neutron knock-out from a ^{30}Mg beam at 85.8(15) MeV/u [22].
- Multi-nucleon transfer reactions: $^{26}\mathrm{Mg}(^{11}\mathrm{B},^{8}\mathrm{B})^{29}\mathrm{Mg}$ [23], $^{26}\mathrm{Mg}(^{18}\mathrm{O},^{15}\mathrm{O})^{29}\mathrm{Mg}$ [24], and $^{30}\mathrm{Si}(^{13}\mathrm{C},^{14}\mathrm{O})^{29}\mathrm{Mg}$ [25].

However, only the transfer study by Matta et al. [15] and the knock-out reaction study by Terry et al. [22] directly probe single-particle structure. Knock-out reactions probe hole states, and only weakly populate the negative-parity states that are important for investigating the island of inversion.

The current work has also been recently submitted for publication [26], but a more comprehensive overview of these results is reported here.

5.8.2 *Overview of the States of* ^{29}Mg

From the extracted cross sections, angular momenta were assigned for States 1–10 as labelled in Fig. 5.18. A series of DWBA angular distributions were calculated for $\ell = 0, 1, 2$, and 3 transfer to each state. The computer code DWUCK5 [27] was used, which applies the method of Vincent and Fortune to unbound states (see Sect. 3.8). The optical-model parameters used were from Ref. [8] for deuterons and Ref. [28] for protons, discussed in more detail below. The orbital angular momentum quantum number was assigned on the basis of which calculated distribution best reproduced the measured angular distributions. Details of this fitting procedure are in Appendix D.

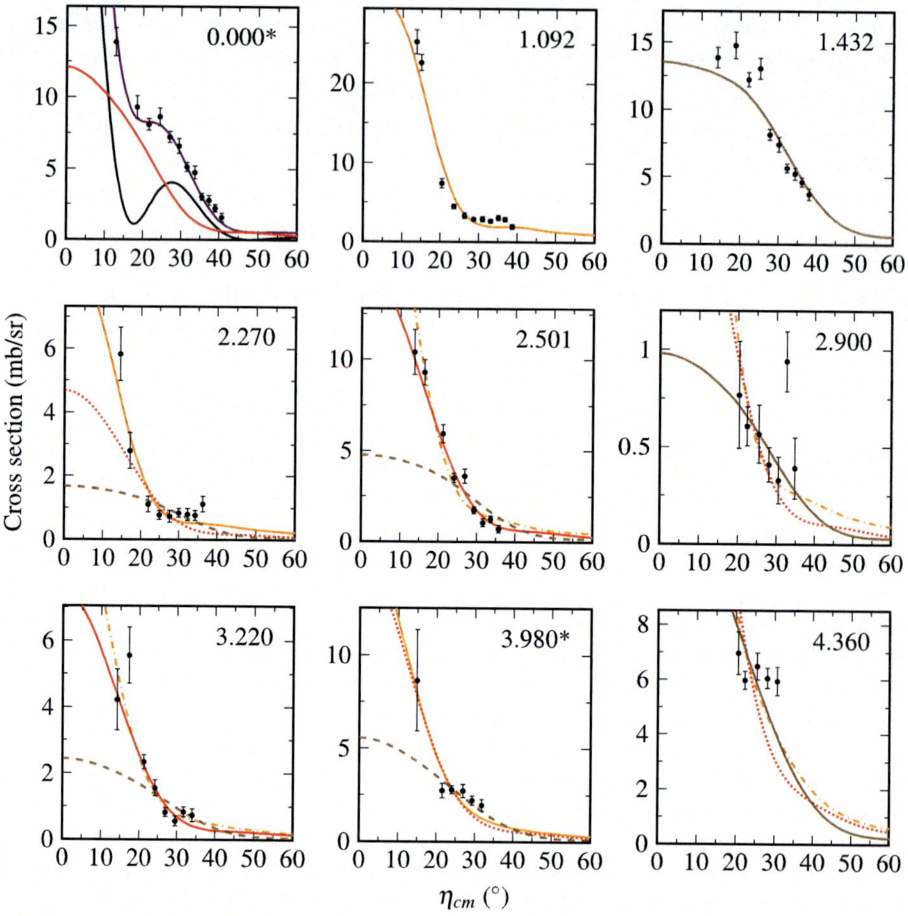

Fig. 5.19 Angular distributions of cross sections for the $d(^{28}Mg,p)^{29}Mg$ experiment, with energies of the states labelled in MeV. The states labelled * are doublets that were not resolved. The angular distributions are coloured according to their value of ℓ: $\ell = 0$ is black, $\ell = 1$ is orange, $\ell = 2$ is red, and $\ell = 3$ is brown; purple corresponds to a sum of two different ℓ values for the ground state ($\ell = 0$ and $\ell = 2$)

From these fits, raw spectroscopic factors were then extracted. These are equal to the scaling factors required to fit the DWBA angular distribution to the data for each state. However, absolute spectroscopic factors are sensitive to the choices in the input parameters and can vary by as much as 50% between different parameter sets. However, their relative values, on which most of the conclusions of this work rely have less variation (10–15%). Here, the raw spectroscopic factors were chosen to be normalised by a procedure similar to methods used in previous work [29].

In this case, the (*d*,*p*) reaction probes the neutron vacancies in the ^{28}Mg core. There are four neutrons expected in the 0*d* orbitals and $1s_{1/2}$ orbital, below the $N = 20$ shell closure. The summed neutron-adding strength (i.e. spectroscopic factors) is related

Table 5.5 The states of ^{29}Mg populated in the $d(^{28}\mathrm{Mg},p)$ reaction

State	E_x (MeV)	η_{cm} (°)	$\frac{d\sigma}{d\Omega}$ (mb/sr)	ℓ	J^{π}	C^2S
1*	0.000	12.8	13.88(100)	2	$\frac{3}{2}^{+}$	0.37(4)
	–	–	–	0	$\frac{1}{2}^{+}$	0.56(3)
2	1.092(3)	13.7	25.23(153)	1	$\frac{3}{2}^{-}$	0.35(2)
3	1.432(2)	14.2	13.86(76)	3	$\frac{7}{2}^{-}$	0.43(2)
4	2.270(18)	14.5	5.82(84)	1	$\frac{1}{2}^{-}$	0.17(1)
5	2.501(6)	13.7	10.41(123)	2	$\frac{3}{2}^{+}$	0.24(1)
6	2.900(32)	20.3	0.77(27)	1	$\frac{5}{2}^{-}$	0.02(1)
7	3.220(16)	14.2	4.22(92)	2	$\frac{5}{2}^{+}$	0.07(1)
8	3.906(13)	14.9	4.27(37)	1	$\frac{1}{2}^{-}, \frac{3}{2}^{-}$	0.24(5), 0.12(2)
9	4.045(22)	14.9	4.37(36)	1	$\frac{1}{2}^{-}, \frac{3}{2}^{-}$	0.24(5), 0.13(3)
10	4.360(10)	20.6	6.97(77)	3	$\frac{7}{2}^{-}$	0.18(1)
11	5.623(9)	–	–	–	–	–
12	5.811(11)	–	–	–	–	–
13	6.043(11)	–	–	–	–	–

The lowest value of η_{cm} and its corresponding cross section for each state have been listed. Spectroscopic factors for each state are also listed here. States 8 and 9 had ambiguous assignments, so both possible assignments have been listed

to vacancy via the Macfarlane-French sum rules [30], and the raw spectroscopic factors were normalised to produce a neutron vacancy of 4 in the s and d orbitals.

The normalisation parameter, N_{sd}, is given by,

$$N_{sd} = \frac{4}{\sum_{J}(2J+1)C^2S_J} \tag{5.8}$$

for $J^{\pi} = \frac{1}{2}^{+}, \frac{3}{2}^{+}$, and $\frac{5}{2}^{+}$. The cross sections and angular distributions are shown in Fig. 5.19 and summarised in Table 5.5. Detailed cross sections for these states at all measured angles are in Appendix E.

The effect of using singles data to obtain the low-angle data points can be seen by the larger associated error bars, particularly for states with higher excitation energy. Comparisons between this and previous work is summarised in Table 5.6.

The strongest excited states that were populated in ^{29}Mg are the $\ell = 1$ state at 1092 keV, and the $\ell = 3$ state at 1432 keV, confirming assignments made in knock-out reactions [22].

Additionally, there are some newly resolved states at 2270, 2501, 2900, and 3220 keV, with $\ell = 1, 2, 3$, and 2 respectively. In the previous transfer experiment, these states were fitted with a combination of $\ell = 2$ and $\ell = 3$ angular distributions, but were unable to be resolved. The γ rays in the knock-out measurement were

Table 5.6 A comparison of the extracted excitation energies and assigned angular momenta for different reactions populating ^{29}Mg

This work		Transfer [15]		Knock-out [22]	
E_x (MeV)	ℓ	E_x (MeV)	ℓ	E_x (MeV)	ℓ
0.000	2	0.0	0+2	0.000	–
0.055	0			0.055	–
1.092	1	1.2	1+3	1.095	1
1.432	3			1.431	3
–	–	2.4	2+3	1.642	–
2.270	1			–	–
2.501	2			2.500	(2, 3)
–	–			2.615	(0, 1)
2.900	1			–	–
3.220	2			3.232	2
–	–	4.2	3	3.674	–
3.906	1			–	–
4.045	1			–	–
4.360	3			–	–

Data primarily taken from Fig. 7 in Ref. [15] and Table 2 in Ref. [22]

Table 5.7 The minimum reduced χ^2 values for the fits of each angular distribution to the unassigned states in ^{29}Mg

E_x (MeV)	$\chi^2_{\ell=1}$	$\chi^2_{\ell=2}$	$\chi^2_{\ell=3}$
2.270	***2.94***	6.36	6.11
2.501	4.67	***1.79***	19.49
2.900	3.65	4.73	***2.82***
3.220	*1.99*	**2.56**	7.66
3.980	**4.11**	6.54	*1.47*
4.360	11.69	26.04	***10.46***

The assigned angular momentum is in bold, with the minimum χ^2 value italized

able to do better, though only some of these states were resolved and the angular momentum only tentatively assigned. These states have therefore been resolved in the current work for the first time.

The χ^2 values from the fits of their angular distributions can formalise the angular momentum assignments. These have been tabulated for states 4-10 in Table 5.7 corresponding to the fits in Fig. 5.19.

The state at 2270 keV was assigned $\ell = 1$ based on its angular distribution. The state at 2501 keV has been assigned $\ell = 2$ based on its angular distribution and the limits placed on ℓ by the knock-out measurements. The state at 2900 keV has a more tentative assignment of $\ell = 3$ based on the angular distribution alone. It is difficult to determine the angular distribution between the $\ell = 1$ and 2 distributions for the state at 3220 keV. However, a $\ell = 2$ assignment has been adopted from the knock-out measurement.

Six states above the neutron-separation threshold were identified: the first three were assigned angular momenta based on their angular distributions (3906, 4045, and 4360 keV), but the final three were not (5623, 5811, and 6043 keV). The doublet at 3980 keV (from the states at 3906 and 4045 keV) is discussed in Sect. 5.8.4.

The state at 4360 keV has only five data points, through which all of the angular distributions could be fitted, making a definitive assignment impossible from these data alone. However, the $\ell = 3$ assignment agrees with a number of previous measurements, including the transfer measurement [15]. There is no compelling reason to disagree with these assignments, and also agrees with the minimum χ^2 (for this distribution).

Some of the peaks in the spectrum require specific discussion.

5.8.3 Ground-State Doublet (Peak 1*)

Peak 1* in Fig. 5.18 is a doublet state, containing the ground state and first excited state at 54.6 keV. While this has been experimentally verified before at ISOLDE (see, for example, Ref. [31]), it was not resolved here. The total cross section was fitted with a combination of angular distributions. While the $\ell = 2$ ground state has been widely reported in literature, this work also agrees with the assignment in Ref. [19], that the 54.6 keV state has angular momentum $\ell = 0$. The spectroscopic factors for the doublet were extracted using the fitting parameters, normalised using the prescription described earlier.

5.8.4 Unbound Doublet (Peaks 8 and 9)

The peak at 3980 keV has an angular distribution that is consistent with $\ell = 1$, 2, or 3.. However, it is broader than the bound state peaks. The measured width of this peak, Γ_{peak}, was approximately 240 keV, compared with the widths of bound states that are dominated by the detector resolution, Γ_b, of approximately 150 keV. Unbound states are broader since the natural decay width is generally comparable in overall magnitude to the resolution. The contribution to the total width of the peak from the natural decay width can be estimated for different values of angular momentum, and gives some indication of the possible spectroscopic factor. The total width is

$$\Gamma_{\text{peak}} = \sqrt{\Gamma_b^2 + (C^2 S \Gamma_{sp})^2}, \tag{5.9}$$

Table 5.8 Values of the square-well decay width and corresponding spectroscopic factor in ^{29}Mg for a state at 3980 keV with a range of angular momenta

ℓ	Γ_{sp} (keV)	C^2S
0	1870	0.01
1	203	0.85
2	24.5	58.3
3	0.61	95,000

Calculated from Ref. [32], with $\Gamma_b = 150$ keV and $\Gamma_{peak} = 240$ keV. The width $C^2S\Gamma_{sp} = 187$ keV for all of these cases

where Γ_{sp} is the single-particle width, with $C^2S\Gamma_{sp}$ the net width of the unbound state (see Eq. 3.44). Using a square-well approximation, the single-particle width can be calculated[2] at the excitation energy of 3980 keV for a range of angular momenta, from which the corresponding spectroscopic factor can be extracted [32]. These values are listed in Table 5.8.

The listed values give an indication of the magnitude of the spectroscopic factor based on the measured width alone. While the angular distribution was consistent with $\ell = 1$ or $\ell = 2$, both of these suggested spectroscopic factors would contribute too much to their respective total single-particle strength. The $\ell = 2$ contribution is far too large in this case as well, suggesting that the state realistically contains $\ell = 1$ strength alone.

The large width and this over-contribution to the total strength suggests that the peak must contain multiple states. Additionally, it suggests that these are of differing j, because the contribution to one of the $p_{3/2}$ or $p_{1/2}$ orbitals alone would exceed the limit imposed by the Macfarlane-French sum-rules when added to the strength from the bound states.

While two states were able to be resolved using data from all of the detectors, this was not possible when analysing the spectra when divided into angle bins due to low statistics. Therefore, the cross sections for these states were measured as a sum of both states, using the ratio of the two peaks in the summed excitation spectrum to get individual cross sections and spectroscopic factors. The ordering of j's is ambiguous, so both values have been given in Table 5.5.

5.8.5 *Higher-Lying Unbound States (Peaks 11–13)*

Three unbound states were observed at 5623, 5811, and 6043 keV. These were only seen when combining data from all of the detectors. Therefore, angular distributions were unable to be fitted, and direct ℓ assignments were not possible.

[2] This required a radius, which was taken to be $1.4(1 + A^{\frac{1}{3}})$ fm, and a Q-value which was simply the energy above the neutron-separation energy.

Table 5.9 Cross sections integrated over all of the detectors and possible spectroscopic factors for states above 5 MeV in ^{29}Mg

E (keV)	σ_1 (mb)	σ_2 (mb)	$\ell = 1$	$\ell = 1$	$\ell = 3$
			$C^2 S_{1/2^-}$	$C^2 S_{3/2^-}$	$C^2 S_{5/2^-}$
5623	0.176(20)	0.176(29)	0.428(53)	0.227(28)	0.046(6)
5811	0.049(15)	0.054(21)	0.156(48)	0.083(26)	0.010(3)
6043	0.052(13)	0.094(23)	0.283(74)	0.151(39)	0.020(5)

However, spectroscopic factors were extracted for a range of possible angular momenta. Table 5.9 details the possible spectroscopic factors, if these states occupy the higher-energy negative-parity orbitals in the pf-shell. The spectrosopic factors were found using cross sections integrated over all of the detectors at both array positions, and then averaged.

The DWBA calculations that used DWUCK5 with the method by Vincent and Fortune failed for $\ell = 1$. Therefore, an approximation was employed, taking the excitation energy to be 3970 keV, the highest value of the excitation energy before the calculations began to fail. Similar approaches have been taken in some studies of unbound states by other authors (for example, Ref. [33]). The $\ell = 3$ distributions were unaffected by this, so were calculated normally.

However, for all three states, the spectroscopic factors arising from this approach are large, suggesting these states give surprisingly large contributions to the single-particle orbital they occupy. This was tested by comparing the widths predicted on the basis of these spectroscopic factors with those measured [32].

Natural widths were extracted from the measured spectrum by fitting a convolution of a Gaussian with a width corresponding to the instrument resolution and a Lorentzian peak for each state. As there were three unbound states, three Lorentzian peaks were used to form this distribution. The data-analysis package ROOT [34] was used to fit this convolution, which had a form $f(E_x) * g(E_x)$ where

$$f(E_x) = \exp\left(-\frac{E_x^2}{2\sigma^2}\right), \tag{5.10}$$

corresponding to the Gaussian profile, and

$$g(E_x) = B + \sum_{n=1}^{3} \frac{A_n \cdot \frac{w_n^2}{4}}{\frac{w_n^2}{4} + (E_x - \mu_n)^2}, \tag{5.11}$$

corresponding to three Lorentzian peaks with a constant background, B. The width of these Lorentzian peaks are w_n. The amplitudes, A_n, are used to scale the peak to the histogram, with the excitation energies μ_n known from prior measurements. Once fitted, the spectroscopic factor is

$$C^2 S = \frac{w}{\Gamma_{sw}}, \tag{5.12}$$

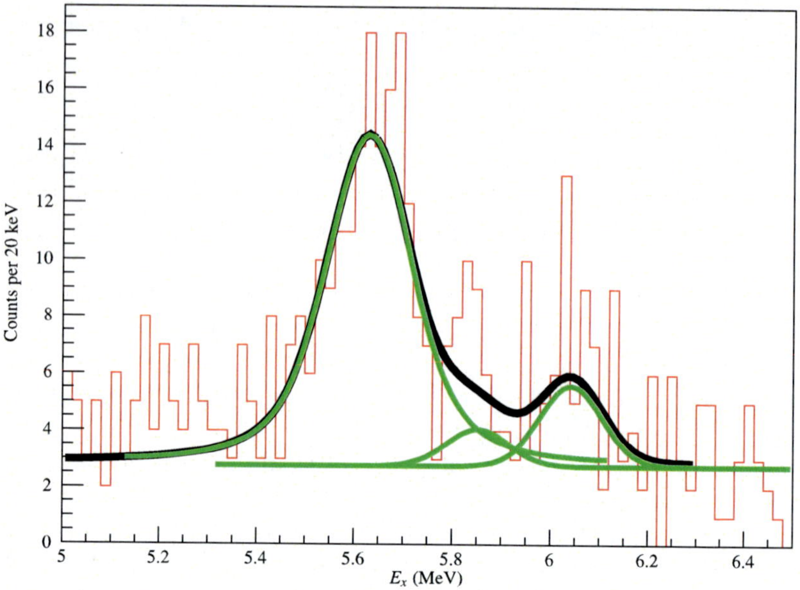

Fig. 5.20 Fits of the three higher-lying excitation states. The profile used is the convolution of the Gaussian and Lorentzian distributions defined in Eqs. 5.10 and 5.11. The total fit is shown in black with individual peaks shown in green

Table 5.10 The widths of the three peaks, and their corresponding spectroscopic factors if the state has $\ell = 1$ or $\ell = 3$

n	w_n (keV)	$C^2S_{\ell=1}$	$C^2S_{\ell=3}$
1	100(50)	0.09(4))	0.5(2)
2	$< 15*$	< 0.01	< 0.06
3	$< 15*$	< 0.01	< 0.05

The asterisks denote states with vanishingly small widths, so an upper limit was estimated as 10% of the bound-state width, 150 keV

where $\Gamma_{sw,n}$ is the width calculated using a square well approximation that implicitly assumes $C^2S = 1$. These fits are shown in Fig. 5.20 and the widths of the states and the corresponding spectroscopic factors are listed in Table 5.10.

Only the first state had a comparable width to the bound states, which have a FWHM of approximately 150 keV. The other two states produced vanishingly small natural widths from the fit. Therefore, an upper limit of 10% of the Gaussian width was used to estimate the natural width.

None of the resulting spectroscopic factors are consistent with those extracted using the DWBA fits to the angular distributions. This suggests that the DWBA calculations could be struggling to reproduce these unbound states at higher excitation, well above the neutron separation energy. At most, therefore, they can be used as an upper limit on the spectroscopic factors. For a proper treatment, more sophisticated calculations would be needed such as those employed by continuum discretisation (see Ref. [35] for an overview). However, as these states do not contain much strength, the effect on the centroids of the single-particle orbitals is not significant, and so don't have a major effect on the physics conclusions of this work.

5.9 Choice of Optical Model and DWBA Code

As mentioned in the previous section, the computer code DWUCK5 has been used for calculating angular distributions that were used for fitting measured cross-sections. A number of different sets of optical-model parameters have been used to assess the variation in *absolute* spectroscopic factors, as detailed in this section, although, as just noted, the main conclusions depend only on their relative numbers.

5.9.1 Comparing DWBA Codes

Calculations of DWBA cross sections depend not only on the choice of optical-model parameters, but potentially on the numerical methods employed by different software codes as well. Checks were made to compare calculations made by a number of different programmes.

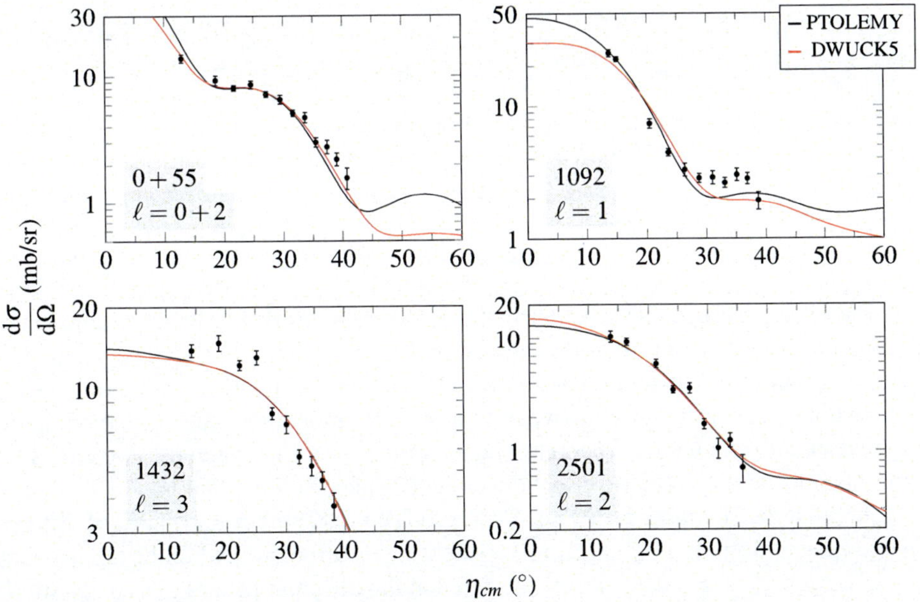

Fig. 5.21 A comparison of the fitted angular distributions using PTOLEMY and DWUCK5. The optical model potentials used are identical for both codes, with the deuteron potential from Ref. [8] and proton potential from Ref. [28]

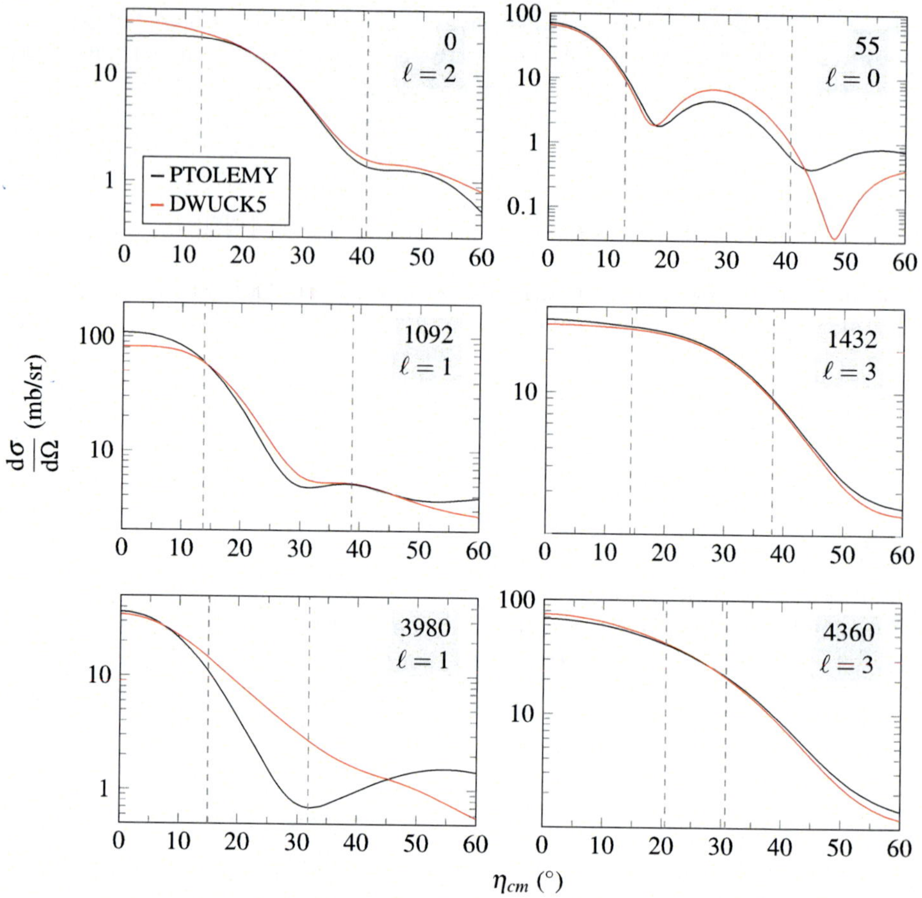

Fig. 5.22 The differences between the raw PTOLEMY and DWUCK5 calculations for a number of states of differing ℓ for ^{29}Mg. The final two plots represent unbound states. The J^{π} for these states are $\frac{3}{2}^{+}$, $\frac{1}{2}^{+}$, $\frac{3}{2}^{-}$, $\frac{7}{2}^{-}$, $\frac{1}{2}^{-}$, and $\frac{7}{2}^{-}$ respectively. The vertical dashed lines represent the angular range where data was obtained for each state

DWUCK5 has been used primarily throughout this work, because it is able to treat unbound states using the method outlined by Vincent and Fortune [36] (see Sect. 3.8). Another commonly used DWBA code is PTOLEMY [7], although the only option for unbound states is to use the weak-binding approximation. These were compared to investigate whether any theoretical uncertainties arise between the two codes that might contribute systematic effects in deduced spectroscopic factors.

Angular distributions for PTOLEMY and DWUCK5 for bound states of differing ℓ have been plotted in Fig. 5.21, to highlight their differences when fitting states.

For each of these states, it can be seen that the different fits are extremely similar when fitted to the angular distributions, particularly over the experimental range of η_{cm}. This shows that both of these codes reproduce the shape of the angular

Table 5.11 The normalised spectroscopic factors for DWUCK5 and PTOLEMY, and their RMS deviation

E_x (keV)	J^π	C^2S		RMS deviation
		PTOLEMY	DWUCK5	
0	$\frac{3}{2}^+$	0.29(6)	0.37(4)	0.08(7)
55	$\frac{1}{2}^+$	0.81(5)	0.56(3)	0.25(6)
1092	$\frac{3}{2}^-$	0.35(2)	0.35(2)	0.004(28)
1432	$\frac{7}{2}^-$	0.35(2)	0.43(2)	0.07(3)
2270	$\frac{1}{2}^-$	0.18(2)	0.17(1)	0.006(21)
2501	$\frac{3}{2}^+$	0.22(2)	0.24(1)	0.03(2)
2900	$\frac{5}{2}^-$	0.021(2)	0.024(2)	0.003(3)
3220	$\frac{5}{2}^+$	0.061(5)	0.071(5)	0.010(7)
3906	$\frac{1}{2}^-$	0.55(12)	0.24(5)	0.3(1)
3906	$\frac{3}{2}^-$	0.29(6)	0.12(2)	0.16(6)
4045	$\frac{1}{2}^-$	0.56(11)	0.24(5)	0.3(1)
4045	$\frac{3}{2}^-$	0.29(6)	0.13(3)	0.17(7)
4360	$\frac{7}{2}^-$	0.153(11)	0.175(9)	0.02(1)

distributions well. Quantifying their difference is more clearly seen by examining the raw outputs of each code, rather than those fitted to data, and this is shown in Fig. 5.22.

Here the angular range of the measured cross sections has been marked with dashed vertical lines, highlighting the regions where differences in the DWBA angular distributions might be important. While most of these distributions look similar in the marked regions, the effect they have on the value of the spectroscopic factors is more important. These values are listed for PTOLEMY and DWUCK5 in Table 5.11.

The RMS deviation between these spectroscopic factors determines how consistent the two spectroscopic factors are, and highlight the effect that the choice of DWBA code has on their values. Most of the bound states have an RMS deviation consistent or almost consistent with zero, showing that the choice of DWBA code for these states is arbitrary.

One notable exception is the 55 keV $\ell = 0$ state. However, examining the two angular distributions in the top right panel of Fig. 5.22 shows that the first maximum is far from the angular range of data, where the differences in the calculations might be exaggerated. This is less of an issue for the other ℓ-values, as their first maxima are closer to or within this angular range.

The unbound states also fare poorly, particularly for the $\ell = 1$ distributions. This is likely due to the failure of the weak-binding approximation in PTOLEMY. The success of the $\ell = 3$ at 4360 keV is probably because the centrifugal barrier is able to contain more of the wavefunction than in the case where $\ell = 1$.

Table 5.12 The optical model parameters used for the ground state $\frac{3}{2}^{+}$ state in ^{29}Mg

	Real			Imaginary			Imaginary surface			Spin orbit			Imaginary spin orbit			R_C	References
	U_v	$r_{0,v}$	a_v	W_v	$\tilde{r}_{0,v}$	$\tilde{a}_v$	W_s	$\tilde{r}_{0,s}$	$\tilde{a}_s$	U_{so}	$r_{0,so}$	a_{so}	W_{so}	$\tilde{r}_{0,so}$	$\tilde{a}_{so}$		
(a)	89.682	1.149	0.758	2.282	1.337	0.548	10.25	1.383	0.721	3.557	0.972	1.011	0	0	0	1.303	[8]
(b)	82.368	1.18	0.744	0	1.27	0.833	10.995	1.27	0.833	6	0.897	0.897	0	0	0	1.3	[9]
(c)	87.011	1.17	0.741	0.448	1.325	0.741	12.245	1.325	0.741	6.781	1.07	0.66	0	0	0	1.3	[10]
(d)	86.075	1.17	0.74	0.444	1.332	0.731	12.144	1.332	0.731	6.594	1.07	0.66	0	0	0	1.3	[10]
(e)	77.188	1.174	0.809	2.186	1.563	0.838	11.981	1.328	0.603	3.703	1.234	0.813	−0.206	1.234	0.813	1.698	[11]
(f)	99.723	1.05	0.86	0	0	0	23.095	1.43	0.623	7	0.75	0.5	0	0	0	1.3	[12]
(g)	84.644	1.15	0.81	0	0	0	18.947	1.34	0.68	0	0	0	0	0	0	1.15	[13]
(A)	53.824	1.17	0.75	1.783	1.32	0.65	9.106	1.32	0.65	6.2	1.01	0.75	0	0	0	1.3	[37]
(B)	52.828	1.173	0.673	1.832	1.173	0.673	8.884	1.293	0.534	5.367	0.977	0.59	−0.094	0.977	0.59	1.315	[28]
(C)	52.242	1.16	0.75	3.034	1.37	0.777	6.281	1.37	0.777	6.04	1.064	0.78	0	0	0	1.25	[38]
(D)	49.824	1.25	0.65	0	0	0	13.5	1.25	0.47	7.5	1.25	0.47	0	0	0	1.25	[39]
(E)	50.99	1.178	0.69	1.75	1.195	0.69	8.667	1.195	0.69	5.9	0.954	0.63	0	0	0	1.279	[40]

Labels with lowercase letters denote deuteron models, and uppercase letters denote proton models. The definition of each symbol is described in Sect. 3.5, with ~ denoting variables that correspond to imaginary potentials

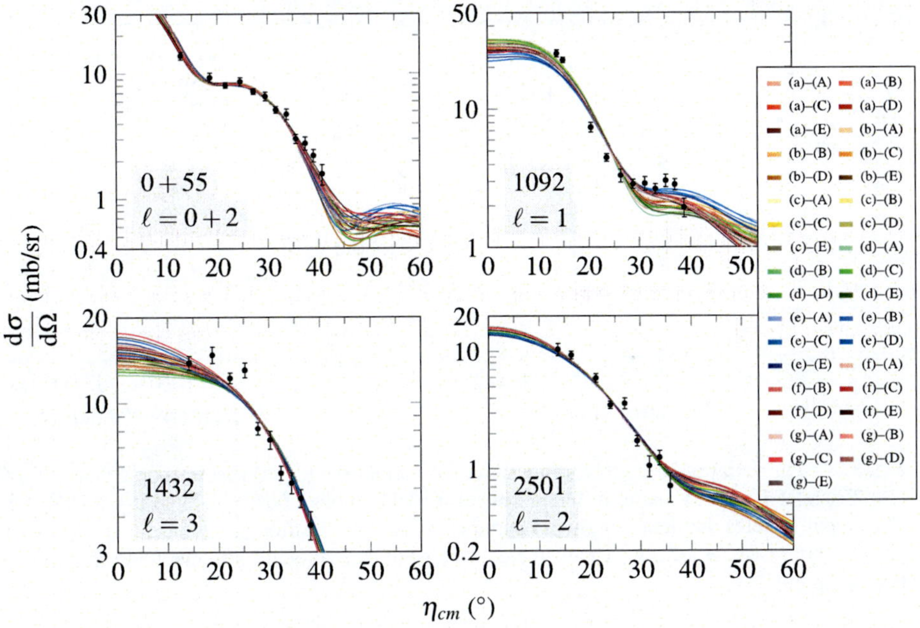

Fig. 5.23 The variation in angular distributions using a number of different optical models to fit cross section data for states in ^{29}Mg. The states are labelled with their energies in keV. The angular distributions were calculated using DWUCK5

Aside from these two caveats, there is little difference between the codes. As DWUCK5 is able to model unbound states using the method of Vincent and Fortune, it is more likely to be a better code to use overall rather than relying on a weak-binding approximation.

5.9.2 Choice of Optical Model

A number of different choices could be made for the optical-model parameters used to model the $d(^{28}\mathrm{Mg},p)^{29}\mathrm{Mg}$ reaction. The global parameters of An and Cai for deuterons [8] and Koning and Delaroche for protons [28] were chosen. In addition to being the best fit to the data, these parameters are consistent with those used in the systematic study in Ref. [29] across a large range of nuclides.

Examining the alternative choices for parameter sets gives an idea of the systematic error arising from the choice of optical model. The parameters used for each optical model are detailed in Table 5.12, and each model is shown in Fig. 5.23 fitted to four states of differing ℓ in ^{29}Mg, calculated using DWUCK5.

Each optical model had parameters derived from fitting elastic-scattering data over a range of energies and masses. This determines their suitability when fitting transfer reactions. Of these models listed, only deuteron models (a), (c), (d), (e), and

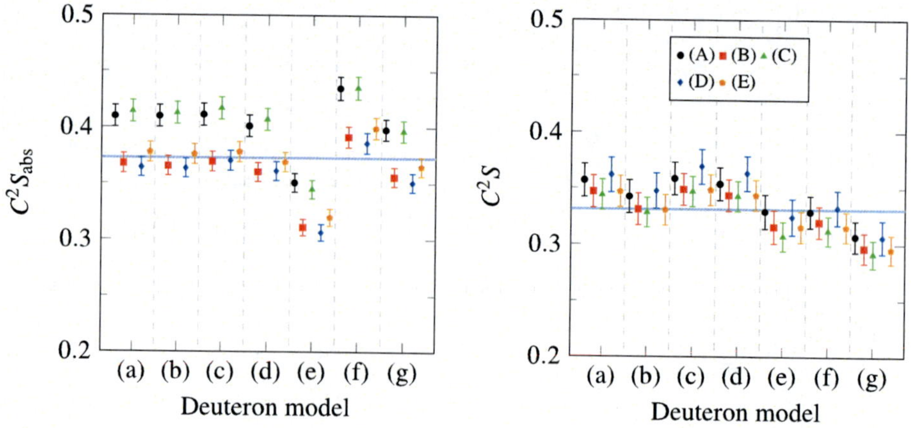

Fig. 5.24 The variation in both absolute (left) and normalised (right) spectroscopic factors for the $\ell = 2$ state at 1092 keV using different optical model combinations calculated using DWUCK5. The colour denotes the deuteron model parameters, and the position along the x-axis denotes the proton model (see Table 5.12). The weighted average of these models is shown for each plot by the horizontal line

Table 5.13 The mean and standard deviation of the normalised and absolute spectroscopic factors for deuteron models (a)–(g) and proton models (A)–(E) calculated using DWUCK5

E (MeV)	0.000	0.055	1.092	1.432	2.270	2.501	2.900	3.220	3.980	4.360
Norm. mean	0.345	0.598	0.331	0.444	0.161	0.243	0.024	0.069	0.456	0.177
Norm. standard deviation (%)	8.528	10.64	6.252	6.018	5.180	3.107	5.116	4.653	5.202	3.925
Abs. mean	0.385	0.683	0.373	0.490	0.182	0.270	0.027	0.078	0.513	0.198
Abs. standard deviation (%)	15.93	10.40	8.494	14.52	8.368	12.33	14.89	7.552	10.03	10.41

They have been normalised according to the convention in Eq. 5.8 using all states up to and including the state at 4.360 MeV. The mean was weighted with the uncertainty in each spectroscopic factor

proton model (B) had parameters that were fitted within the desired ranges of beam energy and nuclear mass. Other models have been plotted for comparison.

Using these different optical models, variations of deduced spectroscopic factors can be calculated. This is shown for the strong singlet state at 1092 keV in Fig. 5.24 for both absolute spectroscopic factors and normalised spectroscopic factors, calculated using Eq. 5.8, showing the variation in absolute and relative magnitude respectively.

The degree of variation is predictably larger for the absolute spectroscopic factors than for the normalised spectroscopic factors. Table 5.13 shows the mean and standard deviation amongst spectroscopic factors extracted using different parameter sets.

This clearly shows that the ratio of the standard deviation to the mean is of the order ±10–15% for the absolute spectroscopic factors, but is lower at approximately ±6% in the relative numbers represented by the normalised spectroscopic factors. However, the ground-state doublet is a notable exception to this, with variations of approximately ±10%. These states, once again, have a greater variation because of the greater uncertainty in the fitting of two distributions. It is also worth noting that the last two columns of Table 5.13 contain three of the unbound states. (The spectroscopic factor used for the unbound doublet at 3.980 MeV was evaluated as if it were one state here.)

References

1. Kinsey RR, Dunford CL, Tuli JK, Burrows TW (2020) The NUDAT/PCNUDAT Program for Nuclear Data. Paper submitted to the 9th International Symposium of Capture Gamma-Ray Spectroscopy and Related Topics, Budapest, Hungary. Oct. 1996. Data extracted from the NUDAT database, version 2.8. https://www.nndc.bnl.gov/nudat2/. Accessed 4 May 2020
2. Shamsuzzoha Basunia M (2012) Nuclear data sheets for A = 29. Nucl Data Sheets 113(4):909–972
3. Lighthall JC (2011) Commissioning of the helical orbit spectrometer: a new device for measuring nuclear reactions in inverse kinematics. PhD Thesis, Western Michigan University. https://scholarworks.wmich.edu/dissertations/432. Accessed 10 May 2019
4. Ziegler JF (2013) SRIM—the stopping and range of ions in matter. www.srim.org. Accessed 25 Jan 2021
5. Gaffney LP. Private communication (unpublished)
6. Cumer B, Beynel A (2018) HIE-ISOLDE—alignment of ISS magnet detectors on XT02 beam line. EDMS Document. Version 1. CERN, Aug 2018. CERN-EDMS: 2021548/1
7. Macfarlane MH, Pieper S. Argonne National Laboratory (unpublished)
8. An H, Cai C (2006) Global deuteron optical model potential for the energy range up to 183 MeV. Phys Rev C 73:054605
9. Bojowald J, Machner H, Nann H, Oelert W, Rogge M, Turek P (1988) Elastic deuteron scattering and optical model parameters at energies up to 100 MeV. Phys Rev C 38:1153–1163
10. Daehnick WW, Childs JD, Vrcelj Z (1980) Global optical model potential for elastic deuteron scattering from 12 to 90 MeV. Phys Rev C 21:2253–2274
11. Han Y, Shi Y, Shen Q (2006) Deuteron global optical model potential for energies up to 200 MeV. Phys Rev C 74:044615
12. Lohr JM, Haeberli W (1974) Elastic scattering of 9–13 MeV vector polarized deuterons. Nucl Phys A 232(2):381–397
13. Perey CM, Perey FG (1963) Deuteron optical-model analysis in the range of 11 to 27 MeV. Phys Rev 132:755–773
14. Radford DC (2000) Notes on the use of the program gf3. https://radware.phy.ornl.gov/gf3/gf3.html. Accessed 1 Oct 2020
15. Matta A, Catford WN, Orr NA, Henderson J, Ruotsalainen P, Hackman G, Garnsworthy AB, Delaunay F, Wilkinson R, Lotay G et al (2019) Shell evolution approaching the N = 20 island of inversion: structure of ^{29}Mg. Phys Rev C 99:044320
16. Diget CA, Fox SP, Smith A, Williams S, Porter-Peden M, Achouri L, Adsley P, Al-Falou H, Austin RA, Ball GC et al (2011) SHARC: silicon highly-segmented array for reactions and Coulex used in conjunction with the TIGRESS γ-ray spectrometer. J Instrum 6(02):P02005–P02005
17. Hackman G, Svensson CE (2014) The TRIUMF-ISAC gamma-ray escape suppressed spectrometer [sic]. TIGRESS. Hyperfine Interact 225(1):241–251
18. Guillemaud-Mueller D, Detraz C, Langevin M, Naulin F, de Saint-Simon M, Thibault C, Touchard F, Epherre M (1984) b-Decay schemes of very neutron-rich sodium isotopes and their descendants. Nucl Phys A 426(1):37–76

19. Baumann P, Dessagne P, Huck A, Klotz G, Knipper A, Miehé C, Ramdane M, Walter G, Marguier G, Giroux J, Richard-Serre C (1987) Gamow-Teller beta decay of ^{29}Na and comparison with shell-model predictions. Phys Rev C 36:765–773
20. Baumann P, Dessagne P, Huck A, Klotz G, Knipper A, Miehé C, Ramdane M, Walter G, Marguier G, Gabelmann H et al (1989) Beta decay of ^{30}Na: experiment and theory. Phys Rev C 39:626–635
21. Shimoda T, Tajiri K, Kura K, Odahara A, Suga M, Hirayama Y, Imai N, Miyatake H, Pearson M, Levy CD et al (2014) Nuclear structure explored by b-delayed decay spectroscopy of spin-polarized radioactive nuclei at TRIUMF ISAC-1. Intruder configurations in ^{29}Mg and ^{30}Mg, the nuclei in the region of the island of inversion. Hyperfine Interact 225:183–191
22. Terry JR, Brown BA, Campbell CM, Cook JM, Davies AD, Dinca D-C, Gade A, Glasmacher T, Hansen PG, Sherrill BM et al (2008) Single-neutron knockout from intermediate energy beams of ^{30,32}Mg: mapping the transition into the "island of inversion". Phys Rev C 77:014316
23. Scott DK, Harvey BG, Hendrie DL, Kraus L, Maguire CF, Mahoney J, Terrien Y, Yagi K (1974) Spectroscopy of exotic nuclei using heavy-ion transfer reactions. Phys Rev Lett 33:1343–1346
24. Fifield LK, Drumm PV, Hotchkis MAC, Ophel TR, Woods CL (1985) The $^{26}Mg(^{18}O,^{17}F)^{27}Na$ and $^{26}Mg(^{18}O,^{15}O)^{29}Mg$ reactions and the level schemes of ^{27}Na and ^{29}Mg. Nucl Phys A 437(1):141–166
25. Woods CL, Catford WN, Fifield LK, Orr NA, Sadleir RJ (1988) Studies of the Tz = 52 nuclei ^{31}Al and ^{29}Mg. Nucl Phys A 476(2):392–412
26. Macgregor PT, Sharp DK, Freeman SJ, Hoffman CR, Kay BP, Tang TL, Gaffney LP, Baader EF, Borge MJG, Butler PA et al (2021) Evolution of single-particle structure near the N = 20 island of inversion. Phys Rev C 104:L051301
27. Kunz PD. DWUCK5. Computer code (unpublished)
28. Koning AJ, Delaroche JP (2003) Local and global nucleon optical models from 1 keV to 200 MeV. Nucl Phys A 713(3):231–310
29. Kay BP, Schiffer JP, Freeman SJ (2013) Quenching of cross sections in nucleon transfer reactions. Phys Rev Lett 111:042502
30. Macfarlane MH, French JB (1960) Stripping reactions and the structure of light and intermediate nuclei. Rev Mod Phys 32:567–691
31. Scheit H, Niedermaier O, Pantea M, Aksouh F, Alvarez C, Ames F, Behrens T, Bildstein V, Boie H, Butler P et al (2004) First results on in-beam g spectroscopy of neutron-rich Na and Mg isotopes at REX-ISOLDE. Proceedings of the sixth international conference on radioactive nuclear beams (RNB6). Nucl Phys A 746:96–102
32. Volya A (2021) Nucracker. A library of web-based programs for nuclear physics and beyond. https://www.volya.net/. Accessed 4 Feb 2021
33. Coker WR (1974) Gamow-state analysis of ^{54}Fe(d, n) to proton resonances in ^{55}Co. Phys Rev C 9:784–786
34. Brun R, Rademakers F (1997) ROOT—an object oriented data analysis framework. Proceedings AIHENP'96 workshop, Lausanne, Sept 1996. Nucl Instrum Methods Phys Res Sect A 389(1–2):81–86. See also "ROOT", Release v6.18/04, 22/09/2019
35. Camacho JG, Moro AM (2014) A pedestrian approach to the theory of transfer reactions: application to weakly-bound and unbound exotic nuclei. In: Scheidenberger C, Pfützner M (eds) The Euroschool on exotic beams, vol IV. Lecture notes in physics, vol 879. Springer, pp 39–66
36. Vincent CM, Fortune HT (1970) New method for distorted-wave analysis of stripping to unbound states. Phys Rev C 2:782–792
37. Becchetti FD, Greenlees GW (1969) Nucleon-nucleus optical-model parameters, A > 40, E < 50 MeV. Phys Rev 182:1190–1209
38. Menet JJH, Gross EE, Malanify JJ, Zucker A (1971) Total-reaction-cross-section measurements for 30–60-MeV protons and the imaginary optical potential. Phys Rev C 4:1114–1129
39. Perey FG (1963) Optical-model analysis of proton elastic scattering in the range of 9 to 22 MeV. Phys Rev 131:745–763
40. Varner RL, Thompson WJ, Mcabee TL, Ludwig EJ, Clegg TB (1991) A global nucleon optical model potential. Phys Rep 201(2):57–119

Chapter 6
Discussion of the $d(^{28}\text{Mg},p)^{29}\text{Mg}$ Experiment and Conclusions

This chapter discusses the results of the $d(^{28}\text{Mg},p)^{29}\text{Mg}$ experiment presented in the previous chapter. Following from the motivations discussed in Chap. 1, the results will be compared to recent and historical shell-model calculations tailored for the region of the nuclear chart around ^{29}Mg. The shell evolution of the $N = 17$ isotones will also be examined. Finally, the future of this particular research will be discussed.

6.1 Comparison to Shell-Model Calculations

6.1.1 Standard Shell-Model Interactions

Historically, in the region of the nuclear chart that includes ^{29}Mg, the presence of intruder configurations has led to issues in theoretical approaches that do not include them in the model space. Established calculations in this region typically fail to reproduce the observed nuclear properties, without the addition of ad-hoc changes. Examples of this include:

- The WBP interaction [1], was modified for use to investigate states in ^{34}P ($N = 19$), produced in the $^{18}\text{O}(^{18}\text{O}, pn)$ reaction, and observed using the γ-rays produced in its de-excitation and β^- decay [2]. The theoretical $0f_{7/2}$ and $1p_{3/2}$ single-particle energies needed to be reduced by 1.8 and 0.5 MeV respectively compared to the original WBP interaction in order to reproduce the 1p–1h states at higher excitation energies.
- The WBP interaction was also modified to reproduce states in ^{30}Al ($N = 17$) and ^{30}Si ($N = 16$) populated in the $^{18}\text{O} + {}^{14}\text{C}$ fusion evaporation reaction in inverse kinematics [3]. The energies of the resulting negative-parity states had to be reduced by approximately 1 MeV relative to the ground state in order to match the experiment.

P. T. MacGregor, *Single-Particle Structure of ^{29}Mg on the Approach to the N=20 Island of Inversion*, Springer Theses, https://doi.org/10.1007/978-3-031-19119-0_6

- The SDPF-M interaction [4] was employed alongside the WBP interaction to analyse states in ^{27}Ne ($N = 17$) produced in the $d(^{26}Ne,p)$ reaction [5]. The SDPF-M interaction was unable to reproduce the energies and the ordering of negative-parity states. The WBP interaction, however, was able to reproduce the ordering of these states, but only when the fp-shell orbital energy had been lowered by 0.7 MeV relative to the sd-shell orbital energy (this shifted variant was labelled WBP-M by the authors).

6.1.2 Shell-Model Interactions Tailored to Cross-Shell Excitations

In recent years, several shell models that account for these cross-shell excitations have been developed for this region, where the model space has been expanded from the sd-shell to include the full pf-shell as well.

SDPF-MU
The SDPF-MU interaction [6], first introduced in Ref. [7], is well established. It has similar foundations to the aforementioned SDPF-M interaction [4], but has differences in a number of areas. While there are some changes in the choice of interactions for different components of the nuclear force (e.g. the chosen pf-shell interaction is GXPF1B [8] rather than Kuo-Brown [9]), the main difference is the addition of a tensor interaction, which was previously discussed in Sect. 2.3.3. The V_{MU} interaction [10] is the foundation of the tensor interaction used in SDPF-MU. It describes the $\pi + \rho$ meson exchange force for this interaction.

The SDPF-MU interaction includes both 0p–0h excitations for positive-parity states and 1p–1h excitations for negative-parity states. This is to match the positive parity sd-shell below the $N = 20$ shell closure and the negative-parity pf-shell above the shell closure. It has been fitted to empirical data to better reproduce the observed energies and ordering of states.

FSU
The FSU interaction [11] is a newer model than the SDPF-MU. It similarly includes 0p–0h and 1p–1h excitations. However, the fitting procedure to derive the interactions has included a greater number of single-particle energies (SPEs) and two-body matrix elements TMBEs in this region, in particular for fp and cross-shell configurations. The model space has also been expanded to encompass the full s-p-sd-fp shells. It uses a number of common components to previously mentioned models. Its base is the WBP interaction [1], but SPEs and TBMEs are replaced in the sd-shell with those of the USDB interaction [12], and in the fp-shell with those of the GXPF1A interaction [13]. More details on this interaction can be found in Refs. [14, 15].

EEdf1
The EEdf1 interaction [16], is based on chiral effective-field theory and the extended Kuo-Krenciglowa method [17] to derive a shell-model interaction. It has been derived

specifically for neutron-rich isotopes of silicon, magnesium and sodium. Again, this uses the *sdpf*-model space, but also includes the effect of three-nucleon forces. This interaction is derived without the need to fit to SPEs or TBMEs.

6.1.3 Comparing Interactions with Experiment

The experimental spectroscopic factors for ^{29}Mg, listed in Table 5.5, are compared to the predicted spectroscopic factors from these shell-model interactions in Fig. 6.1.

Spin assignments in Fig. 6.1 are on the basis of experimental measurements and/or suggestions made by a comparison with the shell-model results (as discussed in detail later). Experimentally, all of the $\ell = 0$ strength is in the first excited state. This is reproduced in all of the models here, though the FSU and SDPF-MU interactions have a few weak $\ell = 0$ states ($C^2S < 0.02$) at higher excitation energies.

The $\ell = 1$ strength was split between four states experimentally. The states at 1092 keV and 2270 keV can be clearly associated with $J^\pi = \frac{3}{2}^-$ and $\frac{1}{2}^-$ respectively by comparison with these calculations. The deduced doublet at 3980 keV matches the shell-model calculations here, with an $\ell = 1$ doublet of differing j near this energy in each calculation. However, the ordering of these J^π remains ambiguous.

The $\ell = 2$ states at 0 and 2501 keV were assigned as $\frac{3}{2}^+$, with the former agreeing with a prior assignment [18]. The state at 3220 keV is weaker, and taken to have $J^\pi = \frac{5}{2}^+$. This state is stronger in the FSU and SDPF-MU calculations than in the EEdf1.

Finally, the $\ell = 3$ states observed at 1432 and 4360 keV correspond well to the strong $J^\pi = \frac{7}{2}^-$ states in all three models shown. The weaker experimental state at 2900 keV also has a $J^\pi = \frac{5}{2}^-$ counterpart in each of these models.

Therefore, there is broad agreement between the modern shell-model calculations and current angular-momentum assignments. The FSU calculation matches the experimental data remarkably well, predicting most of the states near the measured excitation energies (the RMS deviation is 190 keV) and with the assigned angular momenta. The ordering of the 2270 keV $\frac{1}{2}^-$ state and the 2501 keV $\frac{3}{2}^+$ state differ from experiment, but this is the only qualitative difference in relative excitation energies.

The EEdf1 model similarly has analogues to each of the measured states in ^{29}Mg. The ordering of the measured states is identical in this model, and the deviation of the predicted excitation energies from the measured excitation energies is similar to the FSU interaction, with a RMS deviation of 210 keV. Both of these models have a number of weak states ($C^2S < 0.09$) which were not observed by experiment, and are likely below the observational limit of approximately 0.1 mb/sr. The FSU model predicts weak states above the $\frac{7}{2}^-_2$ state, whereas that level is the highest-lying prediction by the EEdf1 calculation.

The SDPF-MU interaction does not match experiment as closely, with an RMS deviation of 350 keV. Where both the FSU and the EEdf1 predict the ground-state doublet states, these are well separated in the SDPF-MU. Additionally, like the FSU

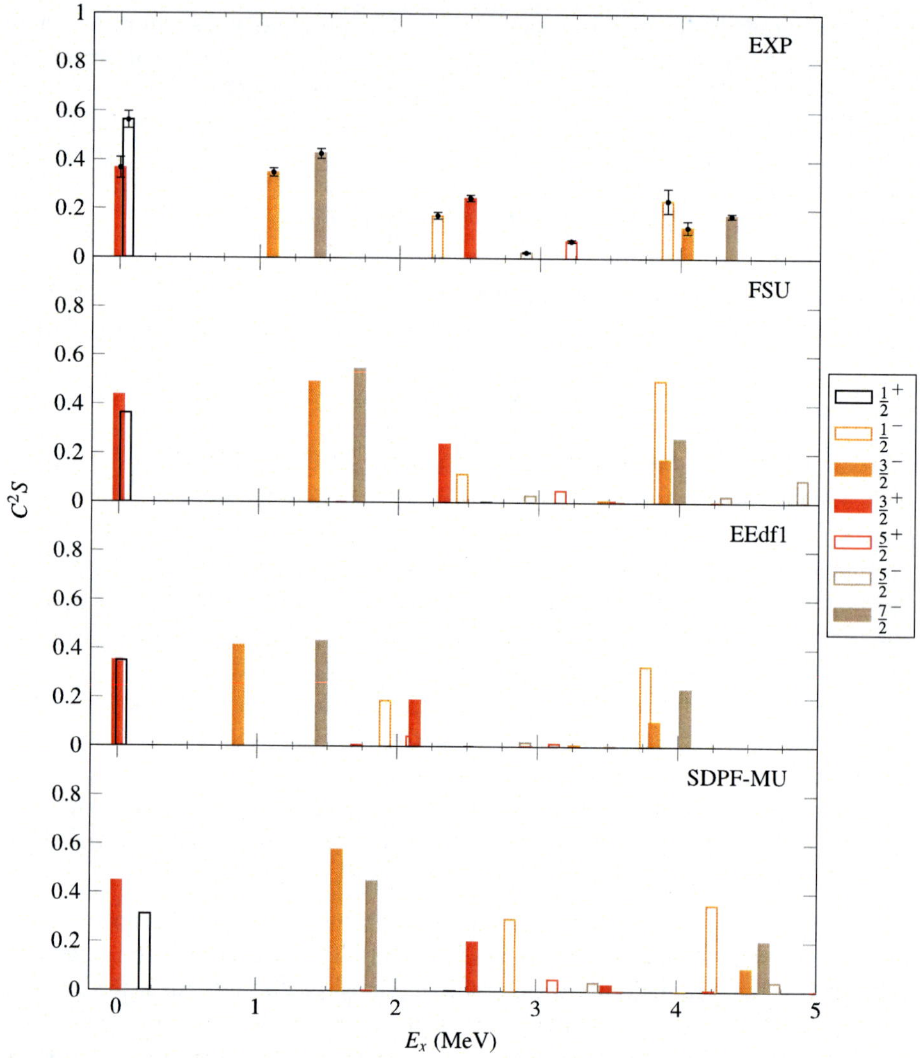

Fig. 6.1 The normalised experimental spectroscopic factors from Table 5.5 compared with predicted states for the FSU, EEdf1, and SDPF-MU models

interaction, it does not reproduce the ordering of the 2270 keV $\frac{1}{2}^-$ state and the 2501 keV $\frac{3}{2}^+$ state. It also predicts more strength in the $\frac{1}{2}^-$ state (calculated at $\approx$ 2800 keV) which was not observed experimentally.

In summary, while there are some minor differences in places over the ordering of levels and exact excitation energies, all these calculations reproduce the states in ^{29}Mg reasonably well, giving confidence to J^π values given on the basis of comparison with experimental states. Therefore, this work improves upon the previous measurements by Matta et al. [19] and Terry et al. [20], by clarifying the energies and angular momenta of the states that lie between the states at 1432 and 4360 keV.

6.2 The $N = 17$ Isotones

6.2.1 Possible Spin Values for Unbound States

The higher unbound states can also be addressed using these calculations. While the EEdf1 calculation does not have higher-lying states above the $\frac{7}{2}^-_2$, the FSU interaction predicts additional strength in the region of the unassigned higher-lying states.

This model, in conjunction with the systematic data, can be used to suggest spin assignments for these higher-lying excited states. Using the summed spectroscopic factors, it is possible to calculate the vacancies of each single-particle orbital in ^{29}Mg, and these are shown in Table 6.1.

Here, these relative vacancies have been compared to experimental results from Piskoř et al. for ^{31}Si [21], and Mermaz et al. and Liljestrand et al. for ^{33}S [22, 23]. As there is ambiguity when assigning spins to $\ell = 1$, the p-orbital vacancies have been combined. Unassigned $\ell = 2$ strength was taken to be $d_{3/2}$, and unassigned $\ell = 3$ strength was taken to be $f_{7/2}$ below 7 MeV and $f_{5/2}$ above.

While these authors report spectroscopic factors, these have been calculated using different reaction codes and optical models, introducing uncertainty when comparing the $N = 17$ isotones. Therefore, the vacancies in Table 6.1 have been calculated using the previously mentioned cross sections and DWUCK5 calculations using the same methodology as for ^{29}Mg. These vacancies have been normalised using the method described in Sect. 5.8.2.

Comparing vacancies across the isotones, there appears to be some strength missing from the p orbitals in ^{29}Mg. While no definitive assignment can be made using the experimental data, the FSU calculations indicate some higher-lying $\ell = 1$ strength, predicting a $\frac{3}{2}^-$ state at 5722 keV, and a $\frac{1}{2}^-$ state at 6079 keV. Assuming that these correspond to the states at 5623 and 5811 keV, this would give a total p orbital vacancy of 3.74(27). The FSU interaction additionally has a state at 6279 keV with $J^\pi = \frac{5}{2}^-$ that corresponds best to the state at 6043 keV. Using these assign-

Table 6.1 Vacancies deduced from the summed spectroscopic factors for the $N = 17$ isotones.

Orbital	$s_{1/2}$	$d_{3/2}$	$d_{5/2}$	p	$f_{7/2}$	Total
^{29}Mg	1.12(7)	2.45(20)	0.43(3)	2.70(17)	4.82(22)	11.5(4)
				3.74(27)		12.6(4)
^{31}Si	0.76(6)	2.91(24)	0.32(3)	3.98(23)	6.40(55)	14.4(7)
^{33}S	0.39(4)	2.99(25)	0.62(4)	4.37(11)	6.13(50)	14.5(6)

The first row of ^{29}Mg corresponds to sums for the states where j has been assigned. The second row includes the strength for the additional unbound states discussed in Sect. 5.8.5. The summed spectroscopic factors for ^{31}Si and ^{33}S come from Refs. [21–23], where spectroscopic factors have been extracted by using DWUCK5 calculations and previously measured cross sections using the same methodology as the current work ments, the total strength in these orbitals has an RMS deviation from the mean of around 7% across the three isotones. These summed vacancies are also reported in Table 6.1.

6.2.2 *Shell Evolution*

The binding energies of the single-particle centroids in ^{29}Mg, ^{31}Si, and ^{33}S are shown in the top panel of Fig. 6.2, illustrating their evolution across the $N = 17$ isotones.

The predictions of the FSU interaction agree well with the measured binding energies of the different single-particle orbitals. The inclusion of the higher-lying unbound states improves agreement with the FSU interaction, particularly for $J^{\pi} = \frac{1}{2}^{-}$, though the $\frac{3}{2}^{-}$ binding energy is higher than the model suggests. However, it has already been established that the DWBA analysis for the higher-lying states gives an upper limit for their respective spectroscopic factors, and hence places an upper limit on the single-particle centroid.

The bottom panels of Fig. 6.2 show the occupancies of positive-parity single-particle orbitals for protons and neutrons in the $N = 17$ isotones, with the latter including experimental results from the present work and Refs. [21–23]. The FSU interaction fits the neutron occupancy data well, though there is currently no available data to corroborate the proton occupancies. Further work in this region using reactions such as (d,^{3}He) would be required to fully verify these predictions.

Nevertheless, the FSU interaction suggests that the protons primarily fill the $\pi d_{5/2}$ shell with increasing Z, so interactions with this orbital must be the main driver of the shell evolution along the isotone. The trends in single-particle energies this drives exemplify Type-I shell evolution (see Sect. 2.3.4).

The $\pi d_{5/2}$ ($j_>$) orbital has a large overlap with the $\nu d_{3/2}$ ($j_<$) orbital, which is greater than those with the $\nu f_{7/2}$ and $\nu p_{3/2}$ orbitals. As the $\pi d_{5/2}$ orbital fills, interactions are therefore strongest with the $\nu d_{3/2}$, but the overall effect is to reduce the energy of each of these orbitals dependent on the size of the overlap.

The tensor part of the interaction acts differently between protons and neutrons. The tensor interaction with the $\pi d_{5/2}$ protons is attractive for the positive-parity $\nu d_{3/2}$ orbital and repulsive for the negative-parity $\nu f_{7/2}$ and $\nu p_{3/2}$ orbitals. The net effect as the $\pi d_{5/2}$ orbital fills is the $N = 20$ shell gap, which occurs as the nuclides in the $N = 17$ isotone become more stable. Conversely, as the $\pi d_{5/2}$ empties, the gaps between the $\nu d_{3/2}$ and higher fp-shell neutrons reduces, affecting the magnitude of the shell closure and the ordering of orbitals.

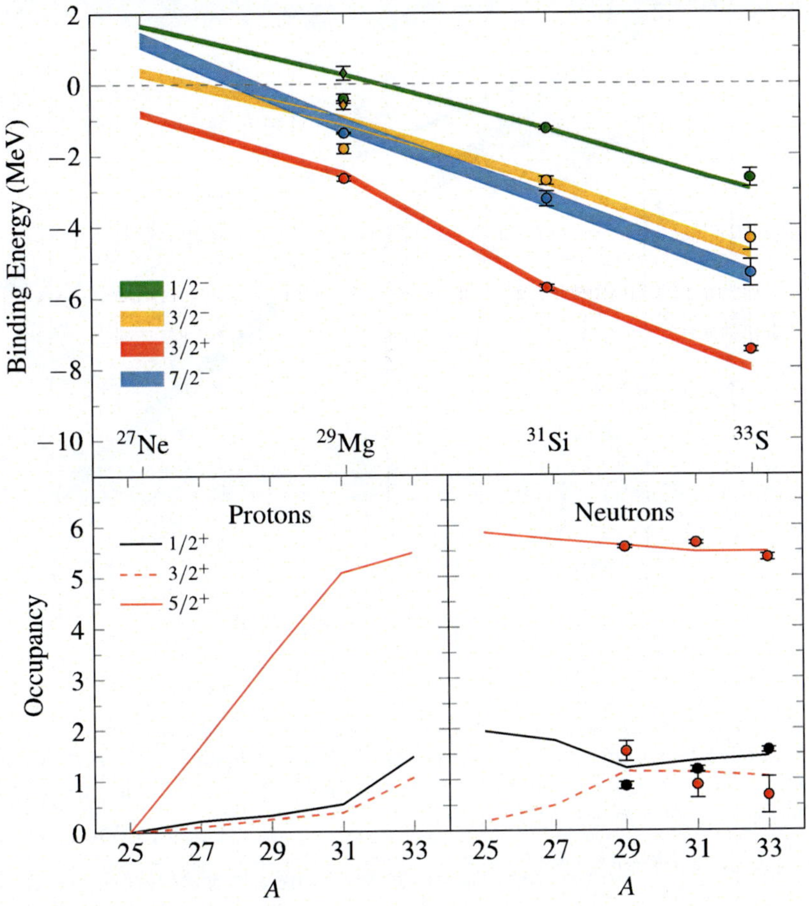

Fig. 6.2 Top: binding energies of the single-particle centroids in the $N = 17$ isotones. Data for ^{31}Si and ^{33}S from Refs. [21–23]. The $\ell = 1$ centroids with states above 5 MeV in ^{29}Mg are denoted with ◇, using DWBA calculations for the spectroscopic factors. Bands denote calculations from the FSU interaction [15] with height corresponding to the uncertainties in ^{31}Si, onto which they are shifted. Bottom: the proton and neutron occupancies for the $N = 17$ isotones from the FSU interaction (lines). Experimental points for the neutrons extracted from the data are also included

6.2.3 *Effects of a Finite Potential*

As single-particle orbitals become less well bound with decreasing Z, the effect of the finite geometry of the potential well plays an increasingly important role, particularly for states with lower ℓ. In other words, the fact that the potential is a finite (rather than infinite) well becomes more apparent in single-particle orbitals close to the top of it. This has been documented in Ref. [24] for p orbitals, where systematics of orbital energies have been investigated for the $N = 21$ isotones. As these orbitals approach the neutron-separation threshold, the rate of change of the energy of these orbitals as a function of radius (or equivalently A) decreases, and they exhibit a "reluctance" to become unbound.

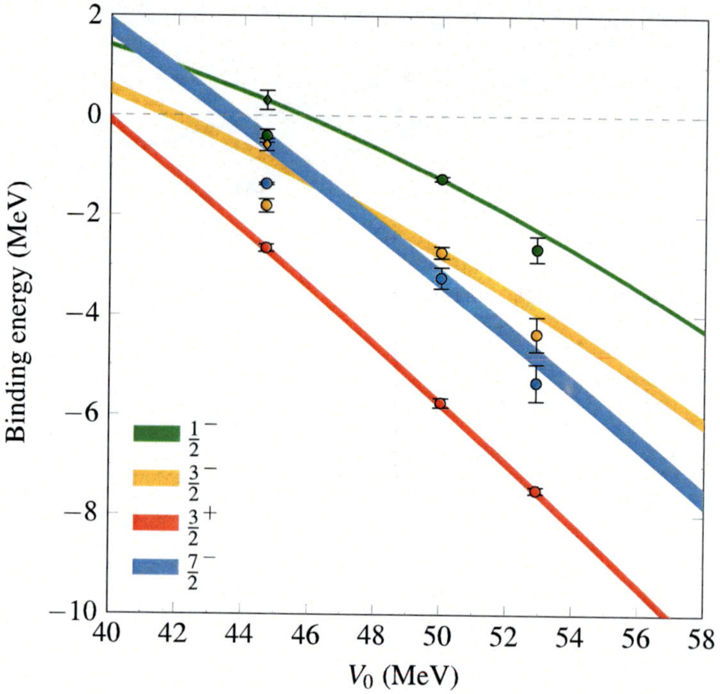

Fig. 6.3 The binding energies of the $N = 17$ isotones with the relative well depth, calculated using a Woods-Saxon formalism with fixed geometry, described in the body text (V_{so} fixed by the calculator). The ^{31}Si experimental data provide uncertainties for the calculated trends. Symbols have the same meaning as in Fig. 6.2

For the $N = 17$ isotones, this was modelled using a Woods-Saxon potential with fixed geometry for $A = 31$ with the computer code of Ref. [25]. Two different geometries were used for this modelling and their results are shown in Figs. 6.3 and 6.4 respectively.

In Fig. 6.3, the geometry was fixed such that $r_0 = 1.2A^{1/3}$ fm, $a_0 = 0.65$ fm, $r_{so} = 1.1A^{1/3}$ fm, $a_{so} = 0.65$ fm. The spin-orbit potential depth proposed by this code using these parameters was $V_{so} = 30.97$ MeV, and was used to produce this figure. The geometry used in Fig. 6.4 was chosen to match the parameters used in Ref. [24], namely $r_0 = 1.28A^{1/3}$ fm, $a_0 = 0.63$ fm, $r_{so} = 1.1A^{1/3}$ fm, $a_{so} = 0.65$ fm, and $V_{so} = 24$ MeV.[1]

In both cases, the depth of the potential well, V_0 was varied to reproduce the binding energy of the $d_{3/2}$ orbital for each isotone, then used to calculate the binding

[1] In Ref. [24], $V_{so} = 6$ MeV, due to a factor of 4 difference between the definition used between these authors and the definition in the code from Ref. [25].

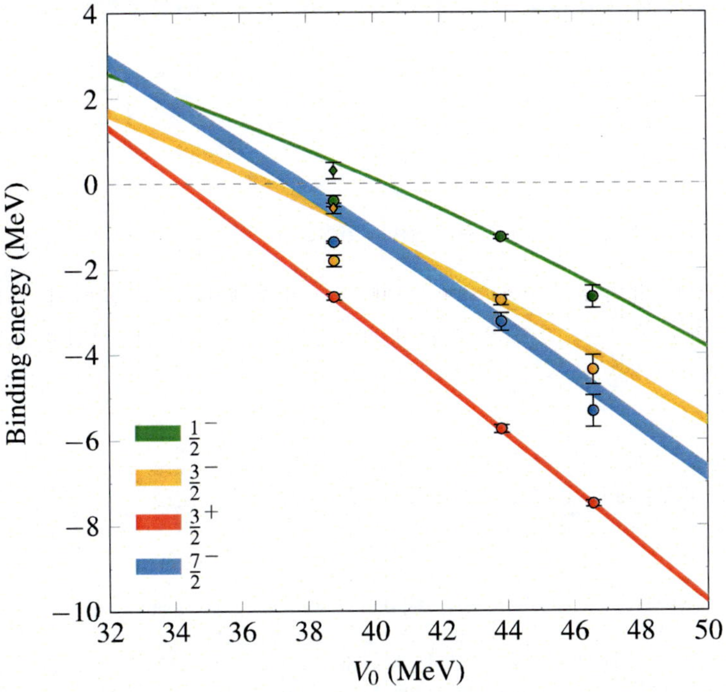

Fig. 6.4 A similar plot as in Fig. 6.3, but using the second set of parameters described in the body text, and where V_{so} has been fixed to match those in Ref. [24]

energy of the other orbitals. The calculations were then shifted onto the experimental energies in ^{31}Si. Uncertainties on these energies generate the width of the theoretical lines in each figure.

The Woods-Saxon calculations reproduce the trends surprisingly well for such a simple model, with the exception of the $f_{7/2}$ orbital in ^{29}Mg. The addition of the higher-lying states makes a significant improvement for the p orbitals. The rate of change of the p orbital energies decreases in both Figs. 6.3 and 6.4, and their separation decreases, as they move closer to the top of the Woods-Saxon potential. This reproduces the lingering effect noted in Ref. [24] in the $N = 21$ isotones.

The effect of changing model parameters has also been investigated, with results shown in Fig. 6.5.

Physically sensible limits were placed on both the potential radius ($1.2A^{1/3} \leqslant r_0 \leqslant 1.3A^{1/3}$) and the diffuseness (0.6 fm $\leqslant a_0, a_{so} \leqslant$ 0.7 fm), and the boundaries of these limits were explored for each orbital. With a similar methodology to that of Figs. 6.3 and 6.4, the calculated trends were fixed to pass through the ^{31}Si points, with the $d_{3/2}$ orbital fixing the value to which the other values of V_0 were shifted. The corresponding values of V_0 for ^{29}Mg and ^{33}S therefore change, and the boundaries of these have been marked with the vertical lines in Fig. 6.5. The term V_{so} was fixed to 24 MeV and the ^{31}Si points were fixed to the value of V_0 in Fig. 6.4 in order to match this particular geometry.

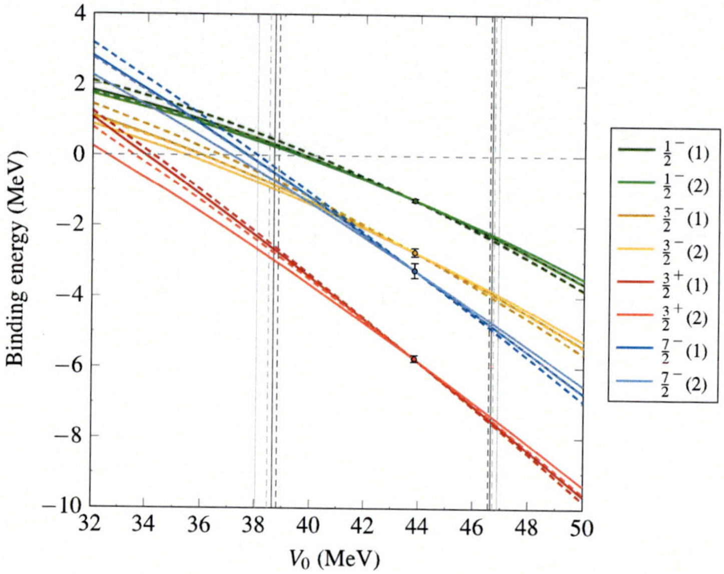

Fig. 6.5 The variation in the relative well depth for the shell-model calculations with differing model parameters fixed to experimental binding energies in ^{31}Si. Colours correspond to the diffuseness of the potentials: darker lines (labelled (1)) have lower diffuseness, and lighter lines (labelled (2)) have higher. Solid lines have lower radii, and dashed lines have higher. The values of V_0 for ^{29}Mg and ^{33}S have been marked with vertical lines using the same colouring. The fitting procedure for these lines is described in the body text

All of the single-particle orbitals show little variation within these limits. Nevertheless, the *p*-orbitals exhibit a similar behaviour irrespective of the model parameters, namely that they have a smaller energy difference as they become more loosely bound and linger below threshold.

6.3 Future Work and Outlook

More work is required to characterise this region of the nuclear chart, in order to improve understanding of the shell evolution that leads to the IoI as well as the weakening of the $N = 20$ shell gap. An obvious extension to this work is to examine more isotones in the $N = 17$ chain, such as the odd-odd ^{30}Al. This was probed in November 2020 at the HELIOS spectrometer using a similar methodology to this work, where a (d,p) reaction was used to populate the single-particle orbitals of ^{30}Al. These data are under analysis.

Of great interest is to extend the current results into the island of inversion by measuring the $^{30}Mg(d,p)$ reaction. With ^{31}Mg inside the island, examining the single-particle evolution that is particularly abrupt along this isotopic chain will be very useful. This was measured at the ISS during the second half of 2021, and is under analysis.

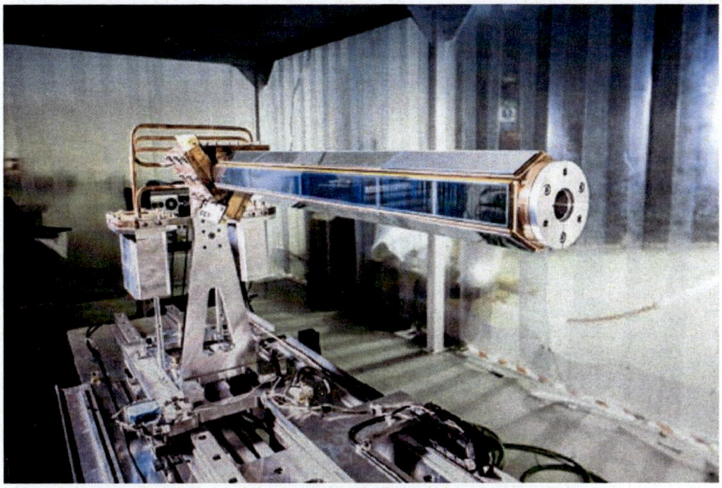

Fig. 6.6 The bespoke silicon array designed for use within ISS [26]

In terms of developing experimental tools, a bespoke array has recently been made for ISS, shown in Fig. 6.6. This has a hexagonal cross section, and uses four double-sided silicon strips per side to measure the position and energy of particles. This is longer than the HELIOS array used in this experiment, allowing for better angular coverage within the detector for a given position. Additionally, it has better energy resolution, and is able to measure position within the chamber to a greater precision.

Transfer reaction studies are continuing at HELIOS at Argonne National Laboratory, but a new solenoidal spectrometer is under construction. The SOLARIS spectrometer [27] has been designed to measure reactions using REA-6 beams available at the Facility for Rare Isotope Beams (FRIB) at Michigan State University (MSU). A schematic diagram of this device is shown in Fig. 6.7.

Two arrays are planned to enable SOLARIS to detect ejectiles in the forwards and backwards hemispheres simultaneously. There is an additional mode of operation using an Active Target Time Projection Chamber (AT-TPC), which has a volume of gas to be used as both a target and detector medium. This allows the tracking of fragments within the chamber. These more advanced methods of detection, coupled with the prolific radioactive beams produced by FRIB promise to advance the exploration of exotic neutron-rich nuclei [27].

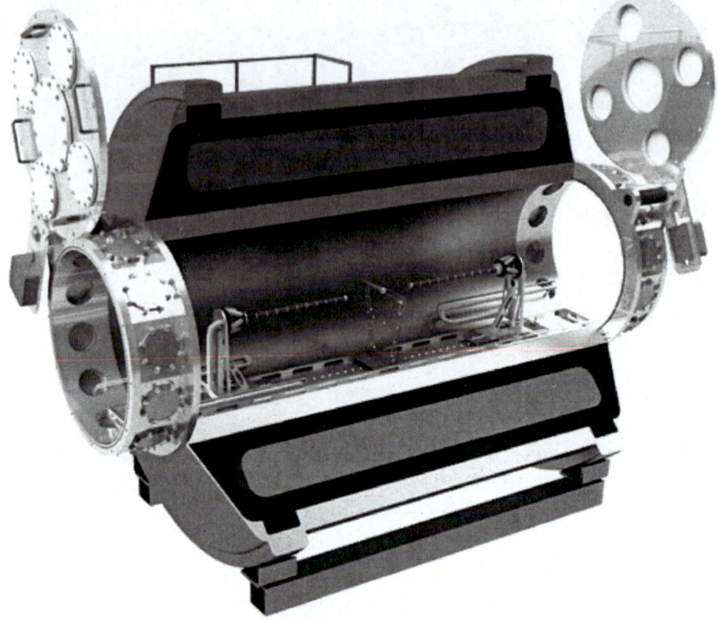

Fig. 6.7 The SOLARIS detector, the newest addition to the family of solenoidal spectrometers [27]

References

1. Warburton EK, Brown BA (1992) Effective interactions for the 0p1s0d nuclear shell-model space. Phys Rev C 46:923–944
2. Bender PC, Hoffman CR, Wiedeking M, Allmond JM, Bernstein LA, Burke JT, Bleuel DL, Clark RM, Fallon P, Goldblum BL et al (2009) Approaching the "island of inversion": ^{34}P. Phys Rev C 80:014302
3. Steppenbeck D, Deacon AN, Freeman SJ, Janssens RVF, Carpenter MP, Hoffman CR, Kay BP, Lauritsen T, Lister CJ, O'Donnell D et al (2010) Cross-shell excitations in ^{30}Al and ^{30}Si at high spin. Nucl Phys A 847(3):149–167
4. Utsuno Y, Otsuka T, Mizusaki T, Honma M (1999) Varying shell gap and deformation in N 20 unstable nuclei studied by the Monte Carlo shell model. Phys Rev C 60:054315
5. Brown SM, Catford WN, Thomas JS, Fernández-Domínguez B, Orr NA, Labiche M, Rejmund M, Achouri NL, Al Falou H, Ashwood NI et al (2012) Low-lying neutron f p-shell intruder states in ^{27}Ne. Phys Rev C 85:011302
6. Utsuno Y (2016) Private communication
7. Utsuno Y, Otsuka T, Brown BA, Honma M, Mizusaki T, Shimizu N (2012) Shape transitions in exotic Si and S isotopes and tensor-force-driven Jahn-Teller effect. Phys Rev C 86:051301
8. Honma M, Otsuka T, Mizusaki T (2008) Shell-model description of neutron-rich Ca isotopes. RIKEN Accel Prog Rep 41:32. https://www.nishina.riken.jp/researcher/APR/Document/ProgressReport_vol_41.pdf. Accessed 7 Apr 2021
9. Kuo TTS, Brown GE (1968) Reaction matrix elements for the 0 f -1p shell nuclei. Nucl Phys A 114(2), 241–279
10. Otsuka T, Suzuki T, Honma M, Utsuno Y, Tsunoda N, Tsukiyama K, Hjorth-Jensen M (2010) Novel features of nuclear forces and shell evolution in exotic nuclei. Phys Rev Lett 104:012501
11. Lubna RS (2019) Private communication

12. Brown BA, Richter WA (2006) New "USD" Hamiltonians for the SD shell. Phys Rev C 74:034315
13. Honma M, Otsuka T, Brown BA, Mizusaki T (2005) Shell-model description of neutron-rich pf-shell nuclei with a new effective interaction GXPF1. Eur Phys J A 25(1), 499–502
14. Lubna RS, Kravvaris K, Tabor SL, Tripathi V, Volya A, Rubino E, Allmond JM, Abromeit B, Baby LT, Hensley TC (2019) Structure of ^{38}Cl and the quest for a comprehensive shell model interaction. Phys Rev C 100:034308
15. Lubna RS, Kravvaris K, Tabor SL, Tripathi V, Rubino E, Volya A (2020) Evolution of the N = 20 and 28 shell gaps and two-particle-two-hole states in the FSU interaction. Phys Rev Res 2:043342
16. Tsunoda N, Otsuka T, Shimizu N, Hjorth-Jensen M, Takayanagi K, Suzuki T (2017) Exotic neutron-rich medium-mass nuclei with realistic nuclear forces. Phys Rev C 95:021304
17. Takayanagi K (2011) Effective interaction in non-degenerate model space. Nucl Phys A 852(1), 61–81
18. Kowalska M, Yordanov DT, Blaum K, Himpe P, Lievens P, Mallion S, Neugart R, Neyens G, Vermeulen N (2008) Nuclear ground-state spins and magnetic moments of ^{27}Mg, ^{29}Mg, and ^{31}Mg. Phys Rev C 77:034307
19. Matta A, Catford WN, Orr NA, Henderson J, Ruotsalainen P, Hackman G, Garnsworthy AB, Delaunay F, Wilkinson R, Lotay G et al (2019) Shell evolution approaching the N 20 island of inversion: structure of ^{29}Mg. Phys Rev C 99:044320
20. Terry JR, Brown BA, Campbell CM, Cook JM, Davies AD, Dinca D-C, Gade A, Glasmacher T, Hansen PG, Sherrill BM et al (2008) Single-neutron knockout from intermediate energy beams of ^{30,32}Mg: mapping the transition into the "island of inversion". Phys Rev C 77:014316
21. Piskoř Š, Novák J, Šimečková E, Cejpek J, Kroha V, Dobeš J, Navrátil P (2000) A study of the ^{30} Si(d, p) ^{31} Si reaction. Nucl Phys A 662(1):112–124
22. Mermaz MC, Whitten CA, Champlin JW, Howard AJ, Bromley DA (1971) Study of the (d,p) reaction on ^{28}Si, ^{32}S, and ^{36}Ar at Ed = 18.00 MeV. Phys Rev C 4:1778–1800
23. Liljestrand R, Mcintyre J, Blanpied G, Lynch J, Ray L, Coker WR, Hoffmann GW (1975) Fractionated single-particle states of ^{33}S at Ex = 8.6–9.5 MeV. Phys Rev C 11:1570–1577
24. Kay BP, Hoffman CR, Macchiavelli AO (2017) Effect of weak binding on the apparent spin-orbit splitting in nuclei. Phys Rev Lett 119:182502
25. Volya A (2021) Nucracker. A library of web-based programs for nuclear physics and beyond. https://www.volya.net/. Accessed 4 Feb 2021
26. Gaffney LP. Private communication (unpublished)
27. Kay BP, Hoffman CR, Wuosmaa AH (2018) SOLARIS: a solenoidal spectrometer apparatus for reaction studies white paper. https://www.anl.gov/sites/www/files/2018-11/solaris_white_paper_final.pdf. Accessed 4 May 2021

Correction to: Single-Particle Structure of ^{29}Mg on the Approach to the $N = 20$ Island of Inversion

Correction to:
P. T. MacGregor, *Single-Particle Structure of ^{29}Mg on the Approach to the N=20 Island of Inversion*, Springer Theses, https://doi.org/10.1007/978-3-031-19119-0

The original version of the book was inadvertently published with an error in the symbol "$\hat{\hbar}$", which has now been corrected as "$\hbar$" in the equations of Chapters 2 and 3. The book has been updated with the changes.

The updated versions of these chapters can be found at
https://doi.org/10.1007/978-3-031-19119-0_2
https://doi.org/10.1007/978-3-031-19119-0_3

P. T. MacGregor, *Single-Particle Structure of ^{29}Mg on the Approach to the N=20 Island of Inversion*, Springer Theses, https://doi.org/10.1007/978-3-031-19119-0_7

Appendix A
Extracting Angles and Excitation Energy in ISS Using Relativistic Kinematics

In Sect. 4.2, a classical derivation of the kinematics for a transfer reaction was presented. This assumed that the angles and excitation energy were known from which z could be extracted. In reality, the situation is reversed, where the ejectile energy and z are measured directly. The kinematics are therefore reformulated to show how the angles and excitation energy are extracted. This is also done relativistically, to reflect how the calculations were performed in the final analysis.

For the following treatment, the value of $c = 1$ for convenience. As with the classical derivation, Figure 4.6 will be instructive. Firstly, the notation will need to be clarified. The Minkowski metric that will be used is

$$[\eta_{\mu\nu}] = \begin{pmatrix} 1 & 0 & 0 & 0 \\ 0 & -1 & 0 & 0 \\ 0 & 0 & -1 & 0 \\ 0 & 0 & 0 & -1 \end{pmatrix}. \tag{A.1}$$

The zeroth component of a four-momentum ${p_i}^{\mu}$ will be written as E_i and the three-momentum will be written as $\mathbf{p}_i$, where i is the number of the particle. The Einstein summation convention will be used. Barred notation will be used to denote that the symbol is in the CM frame instead of the lab frame.

The components of the momentum four-vectors for particles 1 and 2 in the lab frame are

$$[{p_1}^{\mu}] = \begin{pmatrix} E_1 \\ \mathbf{p}_1 \end{pmatrix} = \begin{pmatrix} m_1 + T_1 \\ \mathbf{p}_1 \end{pmatrix}, \tag{A.2}$$

$$[{p_2}^{\mu}] = \begin{pmatrix} E_2 \\ \mathbf{p}_2 \end{pmatrix} = \begin{pmatrix} m_2 \\ \mathbf{0} \end{pmatrix}, \tag{A.3}$$

where T_1 is the total kinetic energy in the lab frame for particle 1, which is the beam energy.

P. T. MacGregor, *Single-Particle Structure of ^{29}Mg on the Approach to the N=20 Island of Inversion*, Springer Theses, https://doi.org/10.1007/978-3-031-19119-0

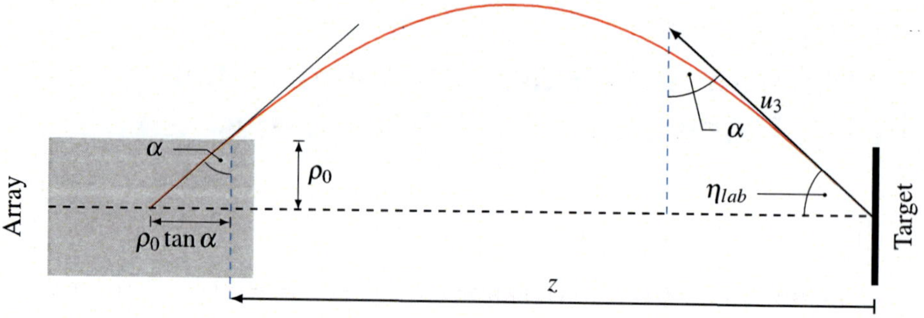

Fig. A.1 A diagram showing the trajectory of a backwards-scattering ejectile following a transfer reaction. The symbols here have been previously defined in Sects. 4.2–4.4. z is negative in this diagram

The total energy in the CM frame, $\overline{E}_{\text{tot}}$, can be calculated with the invariant mass, M. The total invariant mass for the system can be written as

$$M^2 = (p_1 + p_2)^\mu (p_1 + p_2)_\mu = (T_1 + m_1 + m_2)^2 - (\mathbf{p}_1 + \mathbf{0})^2. \tag{A.4}$$

The total momentum four-vector in the CM frame is therefore

$$[\overline{p}_{\text{tot}}{}^\mu] = \begin{pmatrix} \overline{E}_{\text{tot}} \\ \overline{\mathbf{p}}_{\text{tot}} \end{pmatrix} = \begin{pmatrix} M \\ \mathbf{0} \end{pmatrix} = \begin{pmatrix} \sqrt{m_1^2 + m_2^2 + 2(T_1 + m_1)m_2} \\ \mathbf{0} \end{pmatrix}, \tag{A.5}$$

where Eq. A.4 has been simplified using the mass-energy relation

$$E_i^2 = m_i^2 + \mathbf{p}_i \cdot \mathbf{p}_i. \tag{A.6}$$

The gamma-factor for transforming between the lab and CM frames, γ_{cm}, can be found using a Lorentz transformation:

$$E_{\text{tot}} = E_1 + E_2 = \gamma_{cm} \left(\overline{E}_{\text{tot}} + \beta_{cm} \left| \overline{\mathbf{p}}_{\text{tot}} \right| \right) = \gamma_{cm} \overline{E}_{\text{tot}}, \tag{A.7}$$

since there is no total momentum in the CM frame by definition[1]. β_{cm} is the velocity of the frame relative to the speed of light: $\gamma_{cm} = \left(1 - \beta_{cm}^2\right)^{-\frac{1}{2}}$

This has defined the initial system in terms of the CM frame. Once the reaction occurs, the ejectile and residual nucleus will travel antiparallel to one another in the CM frame at an angle, θ_{cm}, with momenta dictated by the excitation energy of the residual nucleus, E_x. In ISS, the ejectile energy and position along the beam axis is then measured. From these measurements, E_x and θ_{cm} can be extracted. Figure A.1 shows the setup for extracting them.

[1] This has already used the fact that β_{cm} and $\mathbf{p}_{\text{tot}}$ will be parallel.

The measured value of z is shorter than the value of z if the array were infinitely thin. The corrected value of z is

$$z = -t_{cyc} \left|\mathbf{u}_3\right| \sin\alpha + \rho_0 \tan\alpha. \tag{A.8}$$

The four-momentum for the ejectile can be expressed as

$$p_3{}^{\mu} = m_3 u_3{}^{\mu} \rightarrow \begin{pmatrix} E_3 \\ \mathbf{p}_3 \end{pmatrix} = \gamma_3 m_3 \begin{pmatrix} 1 \\ \mathbf{u}_3 \end{pmatrix}, \tag{A.9}$$

where γ_3 is the gamma-factor with velocity $\mathbf{u}_3$. Therefore

$$\mathbf{u}_3 = \frac{\mathbf{p}_3}{\gamma_3 m_3}. \tag{A.10}$$

From Eqs. 4.14 and A.10, Eq. A.8 can be rewritten as

$$\frac{\gamma_3 q B}{2\pi} z = -\sqrt{E_3^2 - m_3^2} \sin\alpha + \frac{\gamma_3 q B}{2\pi} \rho_0 \tan\alpha. \tag{A.11}$$

Apart from α, all of these quantities are known. Since this cannot be solved analytically, a Newton-Raphson method can be used to solve for α [1]. Defining a function

$$f(\alpha) = \sqrt{E_3^2 - m_3^2} \sin\alpha - \frac{\gamma_3 q B}{2\pi} \rho_0 \tan\alpha + \frac{\gamma_3 q B}{2\pi} z = 0, \tag{A.12}$$

then α can be iterated using

$$\alpha_{n+1} = \alpha_n - \frac{f(\alpha_n)}{f'(\alpha_n)} \tag{A.13}$$

until $\alpha_{n+1} - \alpha_n$ is less than a sufficiently small threshold. This minimisation yields the value of α. Thus, the ejectile energy in the CM frame is

$$\overline{E}_3 = \gamma_{cm} \left(E_3 - \beta_{cm} \sqrt{E_3^2 - m_3^2} \sin\alpha \right) \tag{A.14}$$

and its momentum is $\left|\overline{\mathbf{p}}_3\right| = \sqrt{\overline{E}_3^2 - m_3^2}$. Therefore, the CM angle can be extracted using the inverse Lorentz transform,

$$E_3 = \gamma_{cm} \left(\overline{E}_3 + \beta_{cm} \left|\overline{\mathbf{p}}_3\right| \cos\theta_{cm} \right) \tag{A.15}$$

$$\Rightarrow \eta_{cm} = \pi - \theta_{cm} = \pi - \arccos\left(\frac{\gamma_{cm} \overline{E}_3 - E_3}{\gamma_{cm} \beta_{cm} \left|\overline{\mathbf{p}}_3\right|} \right). \tag{A.16}$$

After the transfer reaction has taken place, the total momentum in the CM frame can be written as

$$\overline{p}_{tot}{}^{\mu} = \overline{p}_3{}^{\mu} + \overline{p}_4{}^{\mu}. \tag{A.17}$$

The properties of the invariant mass for the residual nucleus can be exploited to give

$$\begin{aligned} \overline{p}_4{}^{\mu}\overline{p}_{4\mu} &= (m_4 + E_x)^2 = \left(\overline{E}_{\text{tot}} - \overline{E}_3\right)^2 - \left(-\overline{\mathbf{p}}_3\right)^2 \\ &= \overline{E}_{\text{tot}}^2 - 2\overline{E}_{\text{tot}}\overline{E}_3 + m_3^2. \end{aligned} \tag{A.18}$$

Therefore, the excitation energy is

$$E_x = -m_4 + \sqrt{\overline{E}_{\text{tot}}^2 - 2\overline{E}_{\text{tot}}\overline{E}_3 + m_3^2}. \tag{A.19}$$

Reference

1. Arfken G (1985) Mathematical methods for physicists, 3rd ed. Academic Press Inc.

Appendix B
The Rate of Change of Centre-of-Momentum Angle with Laboratory Angle

In Sect. 4.3, the claim was made that the term $\frac{d\theta_{cm}}{d\theta_{lab}}$ can be neglected for both NK and IK when looking at its effect upon the derivative $\frac{dT_3}{d\theta_{lab}}$. This is demonstrated more thoroughly here. While the quantity $\frac{d\theta_{cm}}{d\theta_{lab}}$ is the desired derivative, it is easier to first obtain $\frac{d\theta_{lab}}{d\theta_{cm}}$, and then invert it.

Firstly, using Fig. 4.7, the lab angle, θ_{lab}, and CM angle, θ_{cm}, can be related by

$$u_3 \sin\theta_{lab} = v_3 \sin\theta_{cm}, \tag{B.1}$$

$$u_3 \cos\theta_{lab} = v_{cm} + v_3 \cos\theta_{cm}. \tag{B.2}$$

Secondly, the ejectile energy, as defined in Eq. 4.7, is

$$T_3 = \frac{1}{2}m_3 u_3^2 = \frac{1}{2}m_3 v_3^2 + \frac{1}{2}m_3 v_{cm}^2 + m_3 v_3 v_{cm} \cos\theta_{cm}. \tag{B.3}$$

These two combine to give

$$\theta_{lab} = \begin{cases} \arcsin\left(\sqrt{\frac{m_3}{2T_3}} v_3 \sin\theta_{cm}\right), & \theta_{cm} \leqslant \arccos\left(-\frac{v_{cm}}{v_3}\right) \\ \pi - \arcsin\left(\sqrt{\frac{m_3}{2T_3}} v_3 \sin\theta_{cm}\right), & \theta_{cm} > \arccos\left(-\frac{v_{cm}}{v_3}\right). \end{cases} \tag{B.4}$$

θ_{cm} must be defined as a piecewise function because of the limits on the range of the arcsine function, which is $-\frac{\pi}{2} \leqslant \arcsin(x) \leqslant \frac{\pi}{2}$, whereas $0 \leqslant \theta_{cm} \leqslant \pi$. The boundary corresponds to the point where $\theta_{lab} = \frac{\pi}{2}$, and is defined as

$$v_3 \cos\theta_{cm} + v_{cm} = 0 \quad \Rightarrow \quad \theta_{cm} = \arccos\left(-\frac{v_{cm}}{v_3}\right)$$

in terms of θ_{cm}. u_3 has also been written in terms of T_3, because it is a function of θ_{cm} so must be taken into account when taking the derivative.

P. T. MacGregor, *Single-Particle Structure of ^{29}Mg on the Approach to the N=20 Island of Inversion*, Springer Theses, https://doi.org/10.1007/978-3-031-19119-0

Examining the first piece of Eq. B.4, the derivative is

$$\frac{\mathrm{d}\theta_{lab}}{\mathrm{d}\theta_{cm}} = \frac{1}{\sqrt{1-\frac{m_3}{2T_3}v_3^2\sin^2\theta_{cm}}} \cdot \sqrt{\tfrac{m_3}{2}}v_3\left[-\tfrac{1}{2}T_3^{-\frac{3}{2}}\tfrac{\mathrm{d}T_3}{\mathrm{d}\theta_{cm}}\sin\theta_{cm} + T_3^{-\frac{1}{2}}\cos\theta_{cm}\right]. \tag{B.5}$$

Using Eq. B.3,

$$\frac{\mathrm{d}T_3}{\mathrm{d}\theta_{cm}} = -m_3 v_3 v_{cm}\sin\theta_{cm}, \tag{B.6}$$

and using Eqs. B.1 and B.2,

$$u_3^2 = v_3^2 + 2v_3 v_{cm}\cos\theta_{cm} + v_{cm}^2. \tag{B.7}$$

Therefore, Eq. B.5 becomes

$$\begin{aligned}\frac{\mathrm{d}\theta_{lab}}{\mathrm{d}\theta_{cm}} &= \frac{u_3}{v_3\cos\theta_{cm}+v_{cm}}\left[\frac{1}{2T_3u_3}m_3v_3^2v_{cm}\sin^2\theta_{cm} + \frac{v_3\cos\theta_{cm}}{u_3}\right],\\ &= \frac{v_3\left[v_3v_{cm}\left(1+\cos^2\theta_{cm}\right)+\left(v_3^2+v_{cm}^2\right)\cos\theta_{cm}\right]}{\left(v_3^2+v_{cm}^2+2v_3v_{cm}\cos\theta_{cm}\right)\left(v_3\cos\theta_{cm}+v_{cm}\right)}.\end{aligned} \tag{B.8}$$

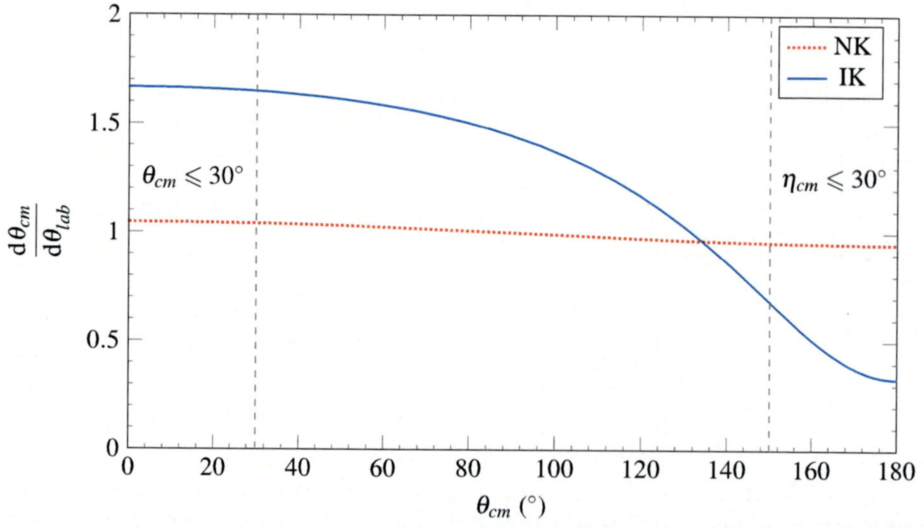

Fig. B.1 The values of $\frac{\mathrm{d}\theta_{cm}}{\mathrm{d}\theta_{lab}}$ for the ^{28}Mg(*d*,*p*) reaction (NK) and *d*(^{28}Mg,*p*) reaction (IK) at $E_x = 0$ MeV. The NK derivative approximately remains constant for all values of θ_{cm} (R=0.048), whereas the IK derivative varies significantly in the desired kinematic region (R=0.669); in this region, $0.331 \leqslant \frac{\mathrm{d}\theta_{cm}}{\mathrm{d}\theta_{lab}} \leqslant 0.686$

This can be rewritten in terms of the ratio $R = \frac{v_{cm}}{v_3}$. The desired derivative is the inverse of Eq. B.8, and is

$$\frac{\mathrm{d}\theta_{cm}}{\mathrm{d}\theta_{lab}} = \frac{\left(1 + R^2 + 2R\cos\theta_{cm}\right)\left(R + \cos\theta_{cm}\right)}{R\left(1 + \cos^2\theta_{cm}\right) + \left(R^2 + 1\right)\cos\theta_{cm}}. \tag{B.9}$$

The desired derivative for the second piece of Eq. B.4 can be derived in a similar manner, though it is the negative of Eq. B.9. However, this is cancelled by the fact that u_3 is also in the opposite direction, so Eq. B.9 can be used for all values of θ_{cm}.

As argued in Sect. 4.2, for addition reactions, $v_3 > v_{cm}$, and so $R < 1$ for the experiments on ^{28}Mg presented in this thesis. For idealised NK, $R \to 0$ as $v_3 \gg v_{cm}$, so $\frac{\mathrm{d}\theta_{cm}}{\mathrm{d}\theta_{lab}} \to 1$. In IK, $R \sim 1$, and so $\frac{\mathrm{d}\theta_{cm}}{\mathrm{d}\theta_{lab}} \to 2$ for most θ_{cm}. However, this function blows up for $\theta_{cm} \to \pi$, which is the angular range which this experiment was examining, so it isn't trivial to determine the effect of R. However, these values have been plotted for the different cases in Fig. B.1.

Although the value of $\frac{\mathrm{d}\theta_{cm}}{\mathrm{d}\theta_{lab}}$ in IK is less than that of NK in the measured angular region, it does not have a significant impact on the magnitude of $\frac{\mathrm{d}T_3}{\mathrm{d}\theta_{lab}}$.

Appendix C
Evolution of Cuts Used in ISS for the $d(^{28}\text{Mg},p)^{29}\text{Mg}$ Experiment

This appendix shows how the cuts in ISS clean the data so that only the desired data remains.

C.1 Evolution of *E* Versus *z* Plots

C.1.1 Raw Data

See Fig. C.1.

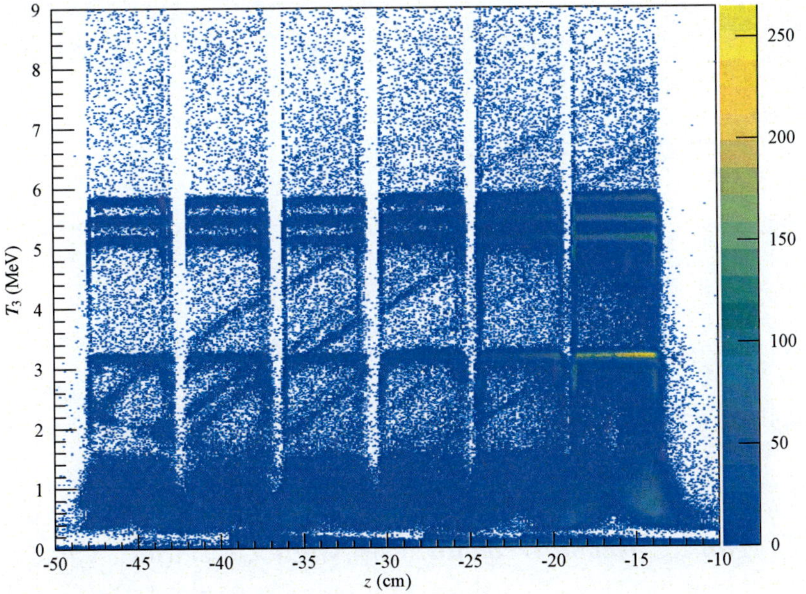

Fig. C.1 T_3 detected on the array with z. No cuts have been applied to the data

P. T. MacGregor, *Single-Particle Structure of ^{29}Mg on the Approach to the N=20 Island of Inversion*, Springer Theses, https://doi.org/10.1007/978-3-031-19119-0

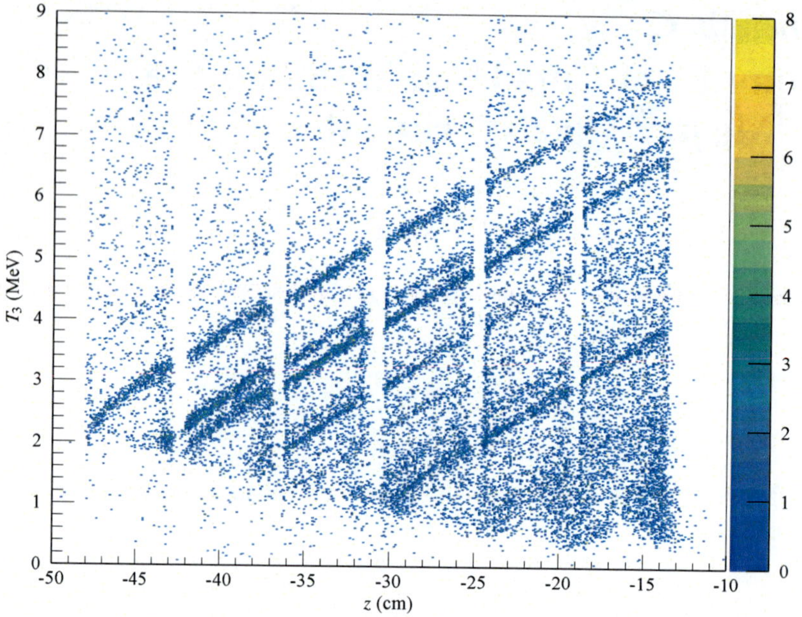

Fig. C.2 T_3 detected on the array with z. Cuts on the time difference between signals on the array and the recoil detector have been applied

C.1.2 *Timing Cuts*

See Fig. C.2.

C.1.3 *Timing and Recoil Detector Cuts*

See Fig. C.3.

C.1.4 *Timing, Recoil Detector, and CM Angle Cuts*

See Fig. C.4.

C.1.5 *Timing, Recoil Detector, CM Angle Cuts, and z Cuts*

See Fig. C.5.

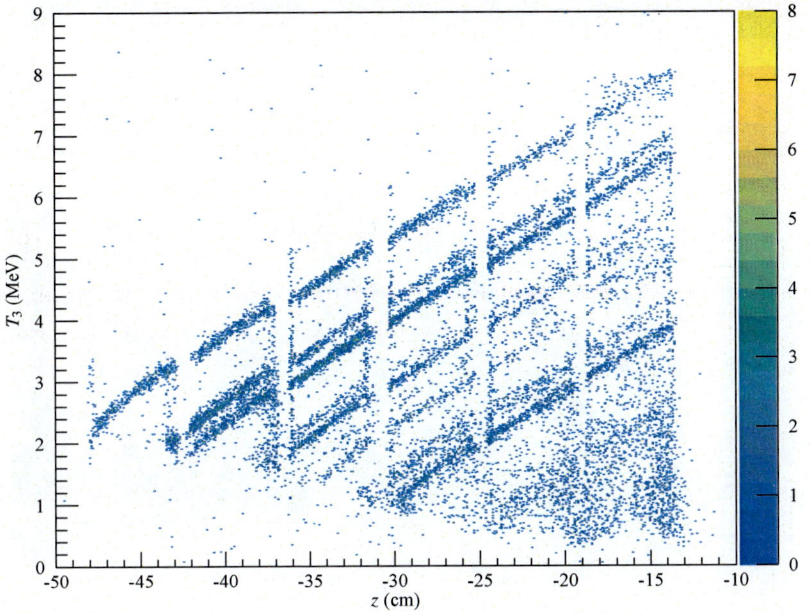

Fig. C.3 T_3 detected on the array with z. This includes the timing cuts from the previous plot, as well as gating on the recoil detector to select for magnesium nuclei

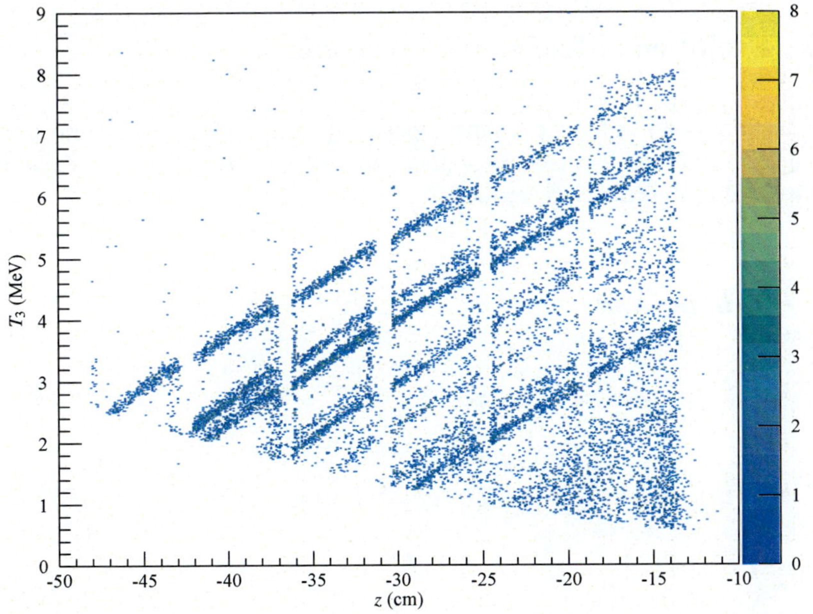

Fig. C.4 T_3 detected on the array with z. This includes the timing cuts and recoil detector gates from the previous plot, as well as cuts that constrain η_{cm}

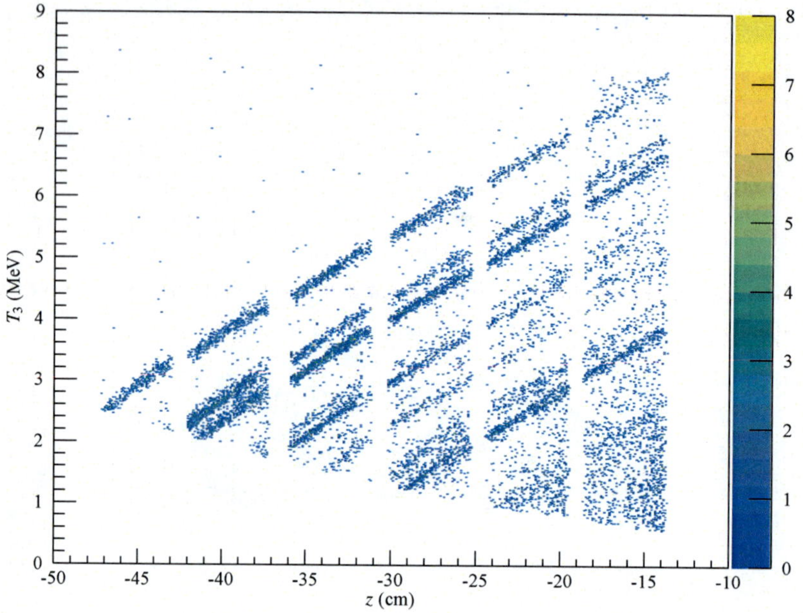

Fig. C.5 T_3 detected on the array with z. This includes the timing cuts, recoil detector gates, and η_{cm} cuts from the previous plot, as well as cuts on z on individual detectors in the array

C.2 Evolution of Excitation Spectrum

This section shows how the excitation spectra change. The grey spectrum is the excitation spectrum before cuts were applied, and the yellow spectrum is the excitation spectrum after the cuts were applied.

C.2.1 Timing Cuts

See Fig. C.6.

C.2.2 Timing and Recoil Detector Cuts

See Fig. C.7.

C.2.3 Timing, Recoil Detector, and CM Angle Cuts

See Fig. C.8.

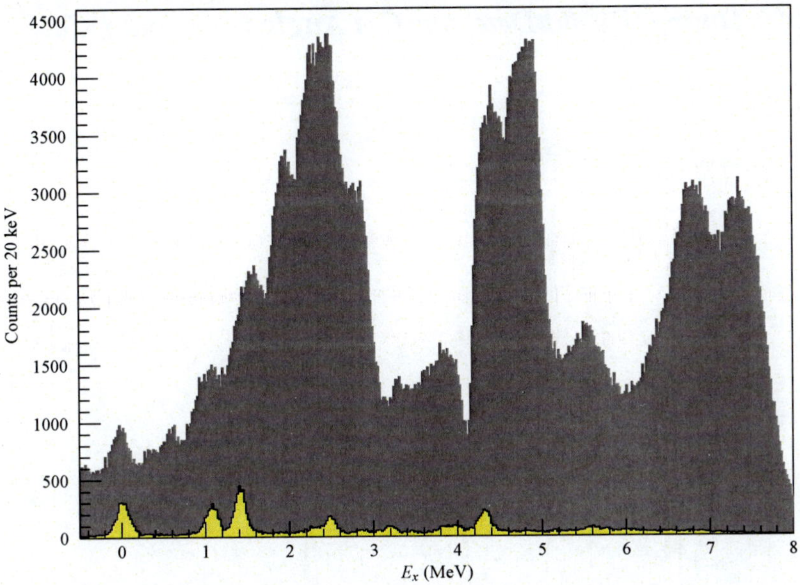

Fig. C.6 The raw excitation spectrum (grey). Applying cuts on the time difference between signals on the array and the recoil detector to this spectrum is shown in yellow

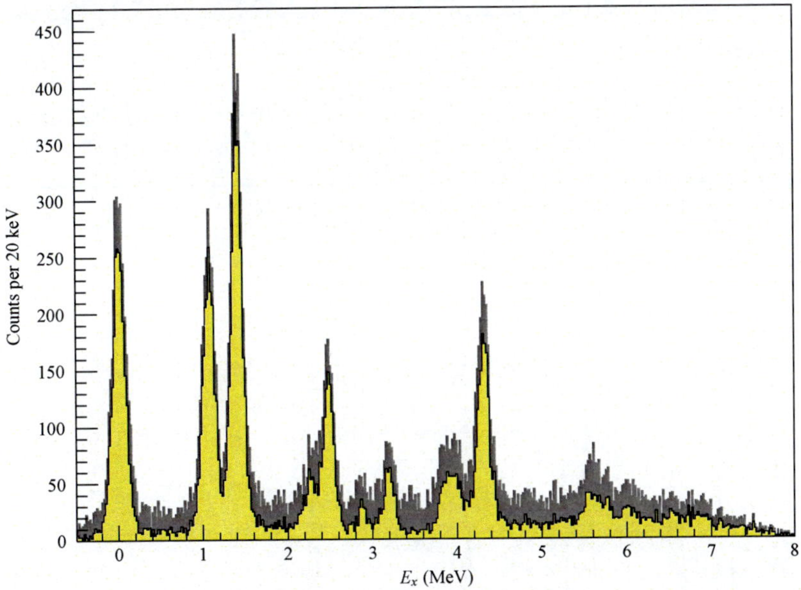

Fig. C.7 The excitation spectrum with timing cuts applied (grey). Applying cuts on the recoil detector to this spectrum is shown in yellow

C.2.4 Timing, Recoil Detector, CM Angle Cuts, and z Cuts

See Fig. C.9.

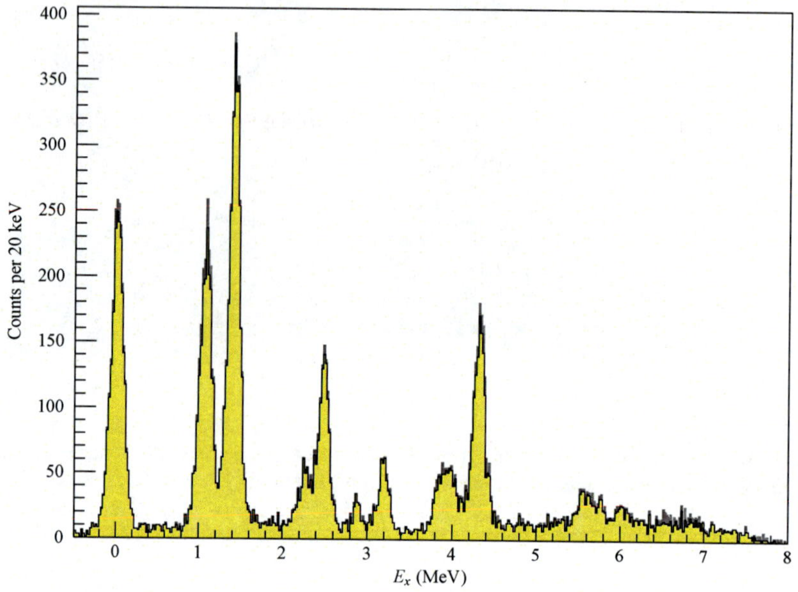

Fig. C.8 The excitation spectrum with timing and recoil detector cuts applied (grey). Applying cuts on η_{cm} to this spectrum is shown in yellow

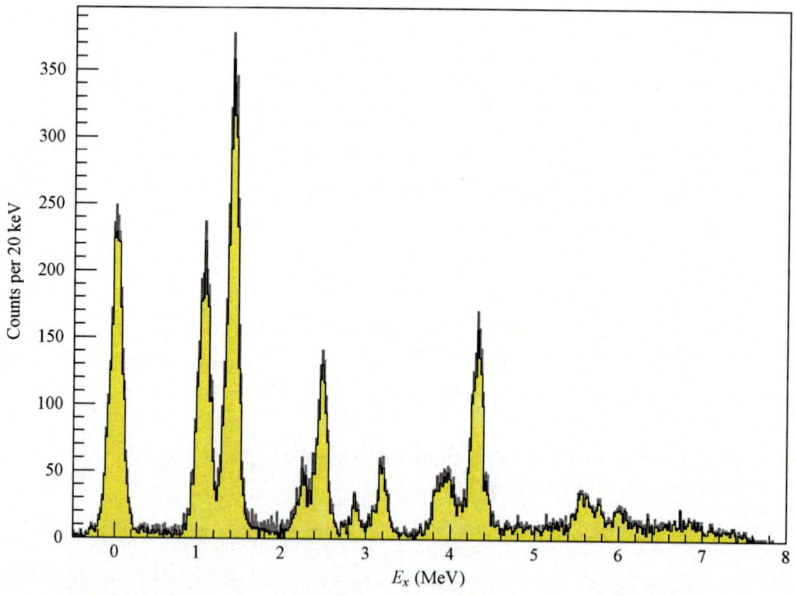

Fig. C.9 The excitation spectrum with timing, recoil detector, and η_{cm} cuts applied (grey). Applying cuts on z to this spectrum is shown in yellow

Appendix D
Fitting Angular Distributions

The method for fitting angular distributions to a series of experimental cross sections is described in the following chapter. The methods used in this chapter have relied on those laid out in Ref. [1]. Some common notation will be used throughout:

- x_i is used for experimental angle i in the data set.
- y_i is used for experimental cross section i in the data set.
- σ_i is used for the statistical error on experimental cross section i in the data set.

D.1 Fitting a Single Angular Distribution

D.1.1 Calculating the Normalisation Factor

Theoretical cross sections can be calculated at each angle x_i to try and reproduce the distribution ω_i. The ω_i will need to be scaled if the state does not carry all of the single-particle strength, so it is necessary to find the best fit of the scaled ω_i to the y_i. In practice, DWBA calculations were used and the normalisation factors were calculated for a series of ℓ-values for the state. To calculate the best fit, the χ^2 can be used, and is defined as

$$\chi^2 = \frac{1}{N} \sum_i \left(\frac{y_i - A\omega_i}{\sigma_i} \right)^2, \tag{D.1}$$

where N is the number of degrees of freedom, and A is the scaling factor for the DWBA distribution. This produces a quadratic polynomial in A with known coefficients. As both y_i and ω_i are positive, this is guaranteed to produce a minimum χ^2 for a particular value of A, which indicates the best statistical fit using this measure. Therefore, the minimum of this χ^2 distribution can be found using

P. T. MacGregor, *Single-Particle Structure of ^{29}Mg on the Approach to the N=20 Island of Inversion*, Springer Theses, https://doi.org/10.1007/978-3-031-19119-0

$$\frac{\partial \chi^2}{\partial A} = 0 = \frac{1}{N}\sum_i 2\left(\frac{y_i - A\omega_i}{\sigma_i}\right)\left(\frac{-\omega_i}{\sigma_i}\right). \tag{D.2}$$

This can be rewritten as

$$\sum_i \frac{y_i\omega_i}{\sigma_i^2} = A\sum_i \left(\frac{\omega_i}{\sigma_i}\right)^2. \tag{D.3}$$

These sums can be rewritten in a simpler way using a barred notation. Defining

$$\overline{Y\Omega} \equiv \sum_i \frac{y_i\omega_i}{\sigma_i^2}, \qquad \overline{\Omega^2} \equiv \sum_i \left(\frac{\omega_i}{\sigma_i}\right)^2,$$

the normalisation factor is therefore

$$A = \frac{\overline{Y\Omega}}{\overline{\Omega^2}}. \tag{D.4}$$

D.1.2 Calculating the Variance

The variance can then be calculated using the formula

$$V(A) = \sum_i \left(\frac{\partial A}{\partial y_i}\right)^2 \sigma_i^2, \tag{D.5}$$

where the only uncertainty stems from the uncertainty in y_i. Substituting Eq. D.4 into Eq. D.5, this can be expressed as

$$V(A) = \sum_i \left(\frac{1}{\overline{\Omega^2}}\frac{\omega_i}{\sigma_i^2}\right)^2 \sigma_i^2. \tag{D.6}$$

Using the same summation notation, this can be shown to be

$$V(A) = \sigma_A^2 = \frac{1}{\overline{\Omega^2}}, \tag{D.7}$$

where σ_A is the uncertainty on A.

D.2 Fitting Two Angular Distributions

The approach outlined in the previous section can be extended to the case of a doublet state, where experimental cross sections are the sum of two angular distributions of differing angular momenta. The maths becomes more involved, but the principles used are the same as in the previous section.

D.2.1 Calculating the Normalisation Factors

As before, a χ^2 can be defined as

$$\chi^2 = \frac{1}{N} \sum_i \left(\frac{y_i - A_\gamma \gamma_i - A_\lambda \lambda_i}{\sigma_i} \right)^2, \tag{D.8}$$

where there are now two normalisation parameters, A_γ and A_λ, for two different theoretical cross sections, γ and λ, which are evaluated once more at each angle x_i to get γ_i and λ_i. It is not as obvious that this χ^2 should have a global minimum. However, the form of the χ^2 is

$$\chi^2 = \kappa_0 A_\gamma^2 + \kappa_1 A_\lambda^2 + \kappa_2 A_\gamma A_\lambda - \kappa_3 A_\gamma - \kappa_4 A_\lambda + \kappa_5 \tag{D.9}$$

where κ_j is a positive constant for $j = 0, \ldots, 5$. This corresponds to the general form of a conic section, showing that the χ^2-surface is either a 2D parabola, ellipse, or hyperbola in 3D space, examples of which are shown in Fig. D.1.

The ellipse and hyperbola each have a critical point, corresponding to a minimum and a saddle point respectively. The parabola has a valley of stability, corresponding to infinitely many minima in the χ^2 surface. The type of critical point can be determined using a Hessian matrix. This is:

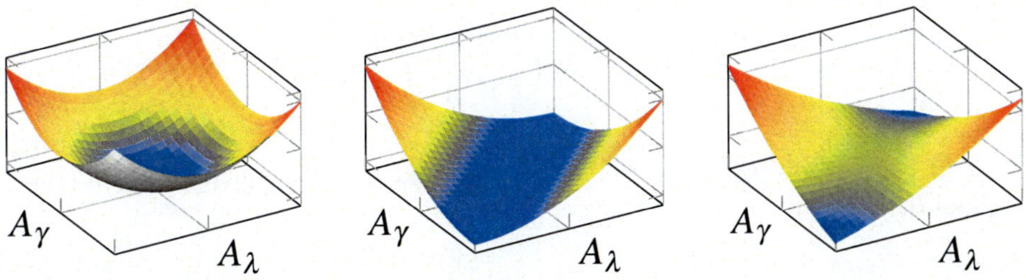

(a) Ellipse (b) Parabola (c) Hyperbola

Fig. D.1 The three different possible χ^2 surfaces for a two-parameter minimisation. The regions of small χ^2 are blue

$$\mathbf{H} = \begin{pmatrix} \dfrac{\partial^2 \chi^2}{\partial A_\gamma^2} & \dfrac{\partial^2 \chi^2}{\partial A_\lambda \partial A_\gamma} \\ \dfrac{\partial^2 \chi^2}{\partial A_\lambda \partial A_\gamma} & \dfrac{\partial^2 \chi^2}{\partial A_\lambda^2} \end{pmatrix} = \begin{pmatrix} 2\kappa_0 & \kappa_2 \\ \kappa_2 & 2\kappa_1 \end{pmatrix}. \tag{D.10}$$

The determinant of this evaluated at the critical points gives information on the outcome [2,3]:

$$\det(\mathbf{H}) = 4\kappa_0\kappa_1 - \kappa_2^2 \Rightarrow \begin{cases} \det(\mathbf{H}) > 0 \Rightarrow 4\kappa_0\kappa_1 > \kappa_2^2 : \text{Ellipse (minimum)} \\ \det(\mathbf{H}) = 0 \Rightarrow 4\kappa_0\kappa_1 = \kappa_2^2 : \text{Parabola (inconclusive)} \\ \det(\mathbf{H}) < 0 \Rightarrow 4\kappa_0\kappa_1 < \kappa_2^2 : \text{Hyperbola (saddle point)} \end{cases}$$

As this determinant does not depend on the values of A_γ or A_λ, it confirms that there is at least one critical point on the surface. The coefficients κ_0, κ_1, and κ_2 can be determined from the derivative of the χ^2 test, which is again set to 0 as a critical point:

$$\frac{\partial \chi^2}{\partial A_\gamma} = 0 = \frac{1}{N} \sum_i 2 \left(\frac{y_i - A_\gamma \gamma_i - A_\lambda \lambda_i}{\sigma_i} \right) \left(-\frac{\gamma_i}{\sigma_i} \right), \tag{D.11}$$

$$\frac{\partial \chi^2}{\partial A_\lambda} = 0 = \frac{1}{N} \sum_i 2 \left(\frac{y_i - A_\gamma \gamma_i - A_\lambda \lambda_i}{\sigma_i} \right) \left(-\frac{\lambda_i}{\sigma_i} \right). \tag{D.12}$$

Using similar bar notation introduced earlier so that

$$\overline{\Gamma Y} = \sum_i \frac{\gamma_i y_i}{\sigma_i^2}, \qquad \overline{\Gamma^2} = \sum_i \left(\frac{\gamma_i}{\sigma_i} \right)^2, \qquad \overline{\Gamma \Lambda} = \sum_i \frac{\gamma_i \lambda_i}{\sigma_i^2},$$

$$\overline{\Lambda Y} = \sum_i \frac{\lambda_i y_i}{\sigma_i^2}, \qquad \overline{\Lambda^2} = \sum_i \left(\frac{\lambda_i}{\sigma_i} \right)^2,$$

then the following matrix equation can be written:

$$\begin{pmatrix} \overline{\Gamma Y} \\ \overline{\Lambda Y} \end{pmatrix} = \begin{pmatrix} \overline{\Gamma^2} & \overline{\Gamma \Lambda} \\ \overline{\Gamma \Lambda} & \overline{\Lambda^2} \end{pmatrix} \begin{pmatrix} A_\gamma \\ A_\lambda \end{pmatrix}. \tag{D.13}$$

Differentiating Eq. D.9 with respect to A_γ and A_λ, and equating coefficients, it can be shown that

$$\kappa_0 = \frac{\overline{\Gamma^2}}{N}, \qquad \kappa_1 = \frac{\overline{\Lambda^2}}{N}, \qquad \kappa_2 = \frac{2\overline{\Gamma \Lambda}}{N}.$$

The determinant can therefore be expressed as

$$\det(\mathbf{H}) = \frac{4}{N^2}\left[\left(\sum_i \frac{\gamma_i^2}{\sigma_i^2}\right)\left(\sum_j \frac{\lambda_j^2}{\sigma_j^2}\right) - \left(\sum_i \frac{\gamma_i\lambda_i}{\sigma_i^2}\right)\left(\sum_j \frac{\gamma_j\lambda_j}{\sigma_j^2}\right)\right], \tag{D.14}$$

which simplifies to

$$\det(\mathbf{H}) = \frac{4}{N^2}\sum_i\sum_j \frac{1}{\sigma_i^2\sigma_j^2}\gamma_i\lambda_j\left(\gamma_i\lambda_j - \gamma_j\lambda_i\right). \tag{D.15}$$

If $i = j$, then the summation is zero. In the case where $i \neq j$, the following notation will prove useful:

$$\gamma_i\lambda_j = \alpha_{ij}; \qquad \gamma_i\lambda_j - \gamma_j\lambda_i = \alpha_{ij} - \alpha_{ji} = \beta_{ij}.$$

This allows Eq. D.15 to be rewritten as

$$\det(\mathbf{H}) = \frac{4}{N^2}\sum_i\sum_{j\neq i} \frac{1}{\sigma_i^2\sigma_j^2}\alpha_{ij}\beta_{ij}. \tag{D.16}$$

As the indices in this sum are dummy variables, they can be swapped freely. By adding this sum to a copy of itself where i and j have been swapped, it is valid to say that

$$\det(\mathbf{H}) = \frac{2}{N^2}\sum_i\sum_{j\neq i} \frac{1}{\sigma_i^2\sigma_j^2}\left(\alpha_{ij}\beta_{ij} + \alpha_{ji}\beta_{ji}\right). \tag{D.17}$$

Using the property that β_{ij} is antisymmetric, then

$$\det(\mathbf{H}) = \frac{2}{N^2}\sum_i\sum_{j\neq i} \frac{1}{\sigma_i^2\sigma_j^2}\left(\alpha_{ij} - \alpha_{ji}\right)\beta_{ij} = \frac{2}{N^2}\sum_i\sum_{j\neq i} \frac{1}{\sigma_i^2\sigma_j^2}\beta_{ij}^2, \tag{D.18}$$

and as this is always positive, it means that the χ^2 surface will always have a minimum.

Therefore, Eq. D.13 can be inverted to produce the following equations for the normalisation parameters which give the minimum point:

$$A_\gamma = \frac{\overline{\Lambda^2}\cdot\overline{\Gamma Y} - \overline{\Gamma\Lambda}\cdot\overline{\Lambda Y}}{\overline{\Gamma^2}\cdot\overline{\Lambda^2} - \left(\overline{\Gamma\Lambda}\right)^2}, \tag{D.19}$$

$$A_\lambda = \frac{\overline{\Gamma^2}\cdot\overline{\Lambda Y} - \overline{\Gamma\Lambda}\cdot\overline{\Gamma Y}}{\overline{\Gamma^2}\cdot\overline{\Lambda^2} - \left(\overline{\Gamma\Lambda}\right)^2}. \tag{D.20}$$

This produces the best possible fit when the two distributions are simply scaled. This is an accurate way of fitting two DWBA distributions to doublet states of differing angular momenta. However, this method finds the best statistical fit, so could produce an unphysical solution where A_γ or A_λ is negative. In this case, numerical methods are required to explore the χ^2-space where A_γ and A_λ are constrained to be positive. If the distributions accurately model the cross section, and the values of j for each state are correct, this should not happen.

D.2.2 Calculating the Variances

As before, the variances can be calculated. However, as there are two normalisation factors, the covariance will also need to be considered. As both A_γ and A_λ have a common denominator, this shall be written as

$$D = \frac{1}{\overline{\Gamma^2} \cdot \overline{\Lambda^2} - \left(\overline{\Gamma\Lambda}\right)^2} \tag{D.21}$$

The variance on A_γ is therefore

$$V(A_\gamma) = \sum_i \left(\frac{\partial A_\gamma}{\partial y_i}\right)^2 \sigma_i^2 = D^2 \sum_i \frac{1}{\sigma_i^4} \left(\overline{\Lambda^2} \gamma_i - \overline{\Gamma\Lambda}\lambda_i\right)^2 \sigma_i^2, \tag{D.22}$$

which can be reduced to

$$V(A_\gamma) = D\overline{\Lambda^2}. \tag{D.23}$$

Similarly,

$$V(A_\lambda) = D\overline{\Gamma^2}. \tag{D.24}$$

The covariance between the two can be calculated using

$$\operatorname{cov}(A_\gamma, A_\lambda) = \sum_i \sum_j \frac{\partial A_\gamma}{\partial y_i} \frac{\partial A_\lambda}{\partial y_j} \operatorname{cov}(y_i, y_j). \tag{D.25}$$

However,

$$\operatorname{cov}(y_i, y_j) = \sigma_i \sigma_j \delta_{ij}, \tag{D.26}$$

where δ_{ij} is the Kronecker delta, and can be used as there is no covariance between points where $i \neq j$.

Following a similar approach to before, this is

$$\operatorname{cov}(A_\gamma, A_\lambda) = -D\overline{\Gamma\Lambda}. \tag{D.27}$$

References

1. Barlow RJ (1989) Statistics: a guide to the use of statistical methods in the physical sciences. Wiley
2. Weisstein EW (2020) Second Derivative Test. From MathWorld–A Wolfram Web Resource. https://mathworld.wolfram.com/SecondDerivativeTest.html. Accessed 2 July 2020
3. Fanchi JR (2006) Math refresher for scientists and engineers, 3rd edn. Wiley

Appendix E
Cross Section Data for ^{29}Mg

This chapter contains the cross section data for states in ^{29}Mg, which is detailed in Table E.1. Each measurement occurred in a different array position for a particular "row" of detectors, denoted Px:Ry for position x and row y. The angles η_{cm} are not directly measured, but are instead calculated on a row-by-row and state-by-state basis.

Table E.1 The complete measured cross sections in ^{29}Mg

	E_x (MeV)	0.000∗	1.092	1.432	2.270	2.501	2.900	3.220	3.980∗	4.360
η_{cm} (°)	P1:R1	17.396	–	–	–	–	–	–	–	–
	P1:R2	21.637	18.684	17.907	–	–	–	–	–	–
	P1:R3	27.138	23.571	22.260	19.804	19.250	–	–	–	–
	P1:R4	31.617	28.755	27.747	24.945	24.074	22.435	21.296	–	–
	P1:R5	35.525	33.104	32.268	30.001	29.314	28.055	26.967	24.030	22.329
	P1:R6	39.002	36.893	36.173	34.245	33.669	32.623	31.732	29.392	28.081
	P2:R1	19.219	–	–	–	–	–	–	–	–
	P2:R2	24.553	20.439	19.684	17.759	–	–	–	–	–
	P2:R3	29.493	26.332	25.201	21.983	21.216	20.328	19.548	–	–
	P2:R4	33.669	31.053	30.142	27.648	26.885	25.473	24.238	21.531	20.615
	P2:R5	37.379	35.132	34.362	32.286	31.663	30.526	29.553	26.970	25.501
	P2:R6	40.716	38.741	38.070	36.280	35.748	34.785	33.968	31.842	30.663
$\frac{d\sigma}{d\Omega}$ (mb/sr)	P1:R1	10.314	–	–	–	–	–	–	–	–
	P1:R2	8.099	10.581	19.613	–	–	–	–	–	–
	P1:R3	7.216	4.477	12.244	1.264	5.883	–	–	–	–
	P1:R4	5.132	2.868	8.163	0.775	3.496	0.607	2.339	–	–
	P1:R5	3.017	2.641	5.673	0.837	1.735	0.409	0.823	2.790	5.977
	P1:R6	2.221	2.844	4.625	0.764	1.244	0.940	0.835	2.214	6.051
	P2:R1	8.086	–	–	–	–	–	–	–	–
	P2:R2	8.627	7.394	12.086	–	–	–	–	–	–
	P2:R3	6.589	3.320	13.071	1.111	5.937	0.767	2.295	–	–
	P2:R4	4.741	2.898	7.433	0.734	3.607	0.567	1.558	2.739	6.966
	P2:R5	2.779	3.041	5.236	0.794	1.060	0.329	0.561	2.723	6.484
	P2:R6	1.579	1.945	3.728	1.118	0.713	0.391	0.735	1.959	5.943

(continued)

P. T. MacGregor, *Single-Particle Structure of ^{29}Mg on the Approach to the N=20 Island of Inversion*, Springer Theses, https://doi.org/10.1007/978-3-031-19119-0

Table E.2 (continued)

$E\left[\frac{d\sigma}{d\Omega}\right]$ (mb/sr)	P1:R1	1.204	–	–	–	–	–	–	–	–
	P1:R2	0.404	0.795	2.954	–	–	–	–	–	–
	P1:R3	0.376	0.284	0.515	0.226	0.556	–	–	–	–
	P1:R4	0.309	0.226	0.405	0.135	0.246	0.102	0.214	–	–
	P1:R5	0.248	0.224	0.328	0.136	0.181	0.088	0.131	0.217	0.326
	P1:R6	0.255	0.270	0.335	0.157	0.177	0.155	0.153	0.244	0.366
	P2:R1	0.614	–	–	–	–	–	–	–	–
	P2:R2	0.619	0.584	0.868	–	–	–	–	–	–
	P2:R3	0.511	0.371	0.777	0.249	0.504	0.275	0.578	–	–
	P2:R4	0.450	0.330	0.549	0.185	0.379	0.150	0.240	0.384	0.772
	P2:R5	0.337	0.329	0.436	0.185	0.214	0.120	0.149	0.353	0.490
	P2:R6	0.302	0.295	0.424	0.242	0.199	0.157	0.195	0.307	0.514

See the body text for further details